Practical Sheep Keeping

For the Small Flock

Practical Sheep Keeping

Kim Cardell

The Crowood Press

First published in 1998 by
The Crowood Press Ltd
Ramsbury, Marlborough
Wiltshire SN8 2HR

www.crowood.com

This impression 2003

**British Library Cataloguing-in-
Publication Data**
A catalogue reference for this book is available
from the British Library

ISBN 1 86126 163 2

Illustrations by Tony Phillips-Smith.
Photographs by Kim Cardell.

Dedication
To my late parents Peggy and Willmar
Cardell, who knew a thing or two about
sheep, and to my brother Peter who
knows even more.

Typefaces used: text, New Century
Schoolbook; labels, Optima.

Typeset and designed by
D & N Publishing
Membury Business Park
Lambourn Woodlands
Hungerford, Berkshire.

Printed and bound by The Bath Press.

Acknowledgements
Grateful thanks to the following friends and col-
eagues who have contributed in so many ways
to the production of this book by reading manu-
scripts, supplying information, and offering
comments, advice and encouragement. They are
all stalwarts of the sheep industry.
Roger Ash; The British Wool Marketing Board;
David Croston, BSc, Head of Sheep Strategy,
Meat and Livestock Commission; Alastair
Dymond; Ron Harrison, BSc; Cath Hoyland,
BSc; Richard Janes, BSc (Animal Sciences);
Peter Johnson; Tony Phillips-Smith; Mike Pret-
tejohn, MRCVS (former president of the Sheep
Veterinary Society); David Sullivan; Ian Wilkin-
son and Robin Hill of Cotswold Seeds.

Contents

Introduction ——————————

Traditionally sheep and Man have a symbiotic relationship – each dependent on the other. Now, in an oil-rich world, Man has become less dependent on sheep but they remain an important element in global agriculture.

Small flocks are vital to the survival of rural communities world wide. This book is an introduction to keeping small flocks. Here, a small flock is around thirty ewes. But, thirty or three hundred, the principles remain the same.

It is for an established farmer seeking to introduce an extra enterprise; for someone with a fascination for sheep and a few acres of land, and for the thousands who come somewhere in between. It fills the gap between those books which introduce us to keeping sheep in the back garden and those which help the large-scale professional sheep farmer to make a better living.

Hopefully it provides a framework of information which allows a small flock – as small flocks are wont to do – to grow into a big flock.

Flock Facts

Life-span	Sheep can live for 20 years.
Breeding span	Average span in a commercial flock is 5 years. More than 10 years in a small flock.
Breeding frequency	Normally once a year. Some breeds capable of breeding every 8 months.
Gestation period	Average 147 days.
Breeding age	Rams able to serve at 4–5 months. Some ewes will breed at 8 months and lamb at 13–14 months.
How many rams	Three adult rams per 100 ewes. One ram lamb per 20 ewes.
Litter size	Hill sheep, one. Lowland sheep, two or three. Four or five not unusual.
Birth weight	Average 4–5kg depending on breed.
Milk production	Dairy ewe 350–500 litres per lactation. Commercial ewe 2 litres per day.
Wool weight	Average 2–4kg fleece per year. 7–10kg from Longwools.
Growth rate	Can average 1.5–2kg per week.
Age fully grown	Usually 18 months to 2 years.
Age of lamb when sold for meat	3–12 months (average 4–8 months) and weighing 35–45kg.
Replacement rate	Average 20 per cent. That is, in a flock of 100 ewes, 20 are culled and replaced each year. Lower in a small flock.
Stocking rate	Excellent grassland 12–14 ewes per hectare for a year. Upland one ewe per hectare. Average 4–5 ewes per hectare.

These figures are averages from a wide range of facts, figures and breeds.

---

1 Where to Begin ———

Sheep are a full-time responsibility and need hands-on attention – much of it physically hard.

A small flock must be established and managed as professionally as a large flock. Not only does commonsense demand it but the law demands it too. Sheep are vulnerable to neglect and ignorance.

The first questions to answer before establishing any flock are:

- Do I have time to keep a daily watch on them?
- Do I have time to sort out problems when they occur?
- Do I have a source of advice?
- Am I physically fit enough to handle sheep?
- Is there help in an emergency?
- Do I have land which is suitable for sheep?
- Can I afford to invest in the basic essentials?
- Can I afford to keep them until they start paying their way?
- Can I accept that they will eventually be sold or slaughtered?

RESOURCES

These questions show that labour, capital, land, equipment and information are the five main resources for sheep keeping.

Labour

The small flock cannot usually afford to pay labour so the owner is the major source, although a flock being established as an extra enterprise on a farm is likely to have labour to call on.

Local contract shepherds will do routine work. Most are trained and experienced shepherds, undertaking fencing, routine health care, dipping, shearing and lambing.

Single-handed shepherding is not difficult with manageable sheep, good fencing, land in one block and good handling pens. For transporting, dipping, shearing and lambing an extra hand may be needed.

In the event of illness or holidays there must be one person who can be called on to check the sheep daily, recognize a problem, be able to deal with it and have the authority to call professional help such as a veterinary surgeon.

As a rule of thumb allow one hour every day to check the flock and deal with any immediate problems and, on average, one day a week for routine management. The lambing period can require at least a month of dedication. Management must be planned to suit other demands on labour. One flock in the UK synchronizes mating and lambing so that the flock lambs over a two-day period at Easter when the owner is home from business.

Capital

Setting-up costs depend on what is already available. If the land is well-fenced and watered and has a suitable building the main expense will be the sheep and basic equipment (Figs 1 and 2).

Fig 1 Basic sheep management equipment includes (from the left): stock fencing, hay rack and bale, torch, buckets, electric fencing components and hurdles.

Budget for everything at the start, even if some items such as lambing equipment are not immediately needed; it may be at least twelve months after buying the sheep before any money comes in.

Setting up on a shoestring is challenging but be prepared to substitute time for money by making equipment, buying it secondhand or moving stock around cheap rented land. But never cut costs on sheep care and always make fencing a priority.

Land

Most small flocks are established on farms where the land is adjacent to the farmhouse. Land which is not within sheep or human walking distance of the home can involve considerable costs in transport, time and inconvenience.

Land for sheep should be well-fenced, well-drained, have a water supply and shelter. If it appears to be suitable but has never carried sheep before it is worth finding out the reason. There may be a mineral deficiency in the soil or a predator problem.

A stream can supply water provided it is accessed safely and legally and stock do not pollute it. Conversely it may already be polluted upstream and this, plus any wet areas, are a health hazard and should be fenced off.

Shelter in the form of trees, hedges or simple housing is essential, and a building, if only for housing a sick animal, is necessary.

Renting Land

In autumn and winter, when cattle are housed, a neighbouring farm may welcome sheep to eat off surplus grass, graze a pasture reseed or even scavenge waste vegetable crops. But beware of basing flock numbers and policy on this supply of grazing because it may not be reliable.

Check that the land is suitable and that the flock will not be forced to stay on muddy ground.

Always have a clear agreement with the landowner as to:

- The start and finish dates.
- Which fields can be used.

- That electric fencing is allowed.
- What public rights of way have to be considered.
- Who is responsible for insurance.
- What responsibilities the landowner will have – such as daily checking, fencing, feeding, watering.
- Payment – amount per month or per head, and when due.
- What else is included in the costs – such as feeding hay or silage.

Another source of grazing is from property owners whose land is too small to farm but too big to garden. This makes a useful area for small groups of sheep such as the rams or ewe lambs.

Sometimes farmers or local authorities want sheep to clear ground, reduce weeds or improve fertility. They offer free grazing on areas such as reservoir banks, playing fields, orchards and woods. Again, bear in mind closeness for shepherding and the need for good fencing.

A local land agent specialising in agricultural matters is a good source of advice on buying, renting or grazing land.

Equipment

Fencing and basic handling equipment are the two most important items to have before buying sheep. The former can be a portable electric fencing system (see Chapter 2) and the latter some lightweight hurdles (Fig 1).

Borrowing or sharing equipment with other flocks is an option but frequently both parties need it at the same time. A lamb weigher and a turnover crate – both expensive items – may warrant sharing.

Information

It is important to get some experience before setting up a flock. Agricultural colleges offer practical shepherding courses and training in specific skills. Spending time observing or helping on a sheep farm is ideal.

Sign on with a local veterinary surgeon – preferably one with a sheep specialist in the practice – and pick his brains. Other sheep producers are a major source of help as are farming organizations – especially ones which have regular local meetings.

There are numerous organizations which can answer queries (see Useful Addresses). Farming magazines give up to date advice and information.

Drug companies, fertilizer and feedstuff manufacturers, seed merchants and equipment makers will answer queries and give advice.

Local and regional agricultural shows involve all aspects of the sheep industry and are exceptionally useful for viewing different breeds and for advice and information. Go with a list of queries to avoid being sidetracked and remember that every breeder believes that his breed is the best.

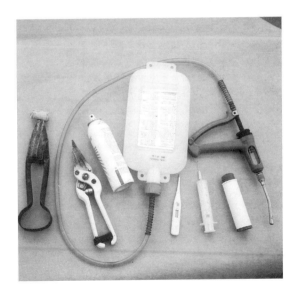

Fig 2 Basic sheep husbandry kit. From the left: wool shears; foot shears; antibiotic spray; worm drench container and gun; digital thermometer; syringe with disposable needle; wax marker for temporary identification.

THE PURPOSE

Sheep are adaptable animals and can be kept for a variety of purposes. Keeping them because you like them is a good enough reason. But the sheep is a working animal and thrives best when it has a job to do.

When deciding on a purpose, bear in mind:

- The environment. Early lamb production is not suited to hill farms.
- The market. Make sure there is one.
- Finance – setting up pedigree or dairy flocks is more expensive than prime lamb commercial flocks.
- Labour and time. Running dry sheep (sheep which are not being bred) for wool needs less than a breeding flock.
- Satisfaction. Pedigree breeding may be more interesting than producing slaughter lambs.

Producing Meat

This is the major reason for keeping sheep. Lambs are easily sold through markets, direct to abattoirs or slaughtered for 'farm gate' sales or home consumption.

The downside can be high marketing costs when selling small numbers and, for some small flock owners, the emotional difficulty in selling lambs for slaughter.

Pedigree Breeding

Pedigree breeding suits the small flock because individual attention to recording, breeding and showing is important. The average size of the pedigree flock in the UK is only 40 ewes.

A pedigree flock is, however, an expensive entry into sheep keeping, requiring good foundation stock, some specialized equipment, extra time and expense for travelling, preparation, showing and advertising. It is said that the work involved in one pedigree ewe is ten times that of a commercial ewe. On the plus side,

once the flock is established and successful it can be quite lucrative.

One way to get established is to set up a cheap commercial (non-pedigree) breeding flock for the experience, then introduce some pedigree ewes and a ram. The pedigree ram can also be used on the commercial ewes to upgrade them and to make better use of him.

Wool

Keeping 'dry' (unbred) sheep is popular for home spinning, selling fleeces or building a cottage industry on woollen products. Young unbred ewes and male castrates, which are not stressed by breeding, give the best fleeces and are simple to manage.

The system could involve a breeding flock of pedigree or purebred specialist wool breeds at the core, with the best offspring kept for wool production and others sold for slaughter. Specialist wools include the fine wool from the Merino, coloured wool from rare breeds such as the Shetland, and lustre wools from the Wensleydale. In the UK, plans for selling wool to any outlet other than the British Wool Marketing Board (*see* Useful Addresses) should be discussed with them.

Dairying

Sheep milk and products such as cheese, yoghurt and ice cream are popular and give added value to the flock. This is an enterprise which requires a large investment in stock, housing and equipment, and must comply with health, hygiene and other regulations. Being a shepherd is not enough; dairying and marketing skills are essential.

A small flock is unlikely to be viable because of the overhead costs but as a trial, some early-weaned ewes could be hand-milked and products made for home consumption to see if expansion is feasible. Ewes' milk freezes successfully and can be frozen and then sold to a manufacturer for processing. The British Milksheep and the Friesland are popular milking breeds. The source of specialist information is the British Sheep Dairying Association (*see* Useful Addresses).

Visitor Attraction

Few people can resist the appeal of ewes and lambs so a small flock attached to a hotel, bed and breakfast or camping establishment will earn bonus points. It is debatable whether the same visitors will appreciate eating home-grown lamb. The flock must be tame, clean and attractive to look at.

The downside is that where the general public are involved there may be mandatory health and safety requirements – especially concerning zoonoses (the transmission of disease between sheep and humans) – and special insurance. The flock programme may have to suit the visitor season – such as lambing just before the season starts.

More ambitious are specialist Sheep Centres at which visitors can see a range of sheep breeds and buy their products. Sheep milking, shearing, spinning, the history of sheep and even sheep racing are featured to give a day out for the family. Travelling exhibitions, visiting schools and other groups to demonstrate the sheep industry, are another option.

Integration

Integrating a flock with another enterprise is a smart way to use sheep. They complement grassland-based enterprises such as beef, will integrate with trees and can maintain healthy wildflower meadows, cliff tops, heath land and orchards. A flock will also utilize low-cost winter housing that can be used for turkey rearing and other purposes for the rest of the year.

Rare Breeds

The sheep industry is littered with breeds that are out of fashion – sometimes for genuine farming reasons and sometimes because of the vigorous marketing of new breeds. Minor, rare and endangered breeds – such as Manx Loghtan, Hebridean and Soay – can be used in most systems to produce meat, breeding stock, wool and for land management. A number of different breeds make an interesting zoo but to conserve them it is better to select one breed.

Many can be flighty and difficult to shepherd and may not adapt to new conditions; but they are hardy, need less attention than modern breeds and are light to handle.

Finishing Store Lambs

Not all slaughter lambs are sold at weaning because they would flood the market and because not all of them are ready to sell. Therefore weaned lambs, especially those born late to hill and upland flocks, are grown very slowly (stored) then fattened for sale in winter and early spring – often on grass and forage crops on lowland farms. This is a natural way to spread marketing and provide fresh lamb for twelve months of the year. Those sold in the following spring are usually called hoggets.

Buying and finishing store lambs is not an attractive enterprise for the small flock. The lambs are usually difficult to handle and the profit is the difference between the buying and selling price minus transport, food, veterinary costs and any deaths. It is an exercise in buying and selling rather than in shepherding.

For established farms they utilize break crops and surplus grass or cereals. Suitable female store lambs make relatively cheap breeding ewes.

Rearing Ewe Lambs

Similar to finishing store lambs, rearing ewe lambs involves buying in young ewes (ewe lambs) and rearing them for a year before selling them for breeding. Again this

Lambing Periods

Lambing periods should be chosen to:

- Suit the market and prices (Fig 103).
- Suit the resources – late lambing does not need housing.
- Match grass growth or other feed to pregnancy and lactation.
- Avoid clashing with other events, such as spring cultivations.
- Suit the weather conditions.
- Suit the breed – some breeds have short breeding seasons.

Autumn Lambing

Usually confined to lowland flocks where grass and other forage crops are available during the winter. Can graze winter grass when dairy cows are housed. Utilizes early lambing breeds such as the Poll Dorset. Lambs grow slowly over the winter on their mothers and are ready for the high priced Easter market. Lambing percentages may be low but so, too, are production costs. The bulk of the work is during the winter.

Early Lambing

Lambs are born in late winter. Often housed for lambing and reared on forage crops and concentrates. Has high feed costs and needs to catch early prices with fast-growing, top-quality lambs. Can stock the dry ewes heavily in the summer and would suit a farm with limited, rough or droughty summer grassland and which grows forage crops and cereals. Bulk of the work is in late winter and early spring. Lambs are sold at 10–16 weeks of age and could be weaned at six weeks and finished indoors on concentrates while ewes are stocked heavily outside. May suit pedigree flocks which need lambs well grown for sales. Suits quick-maturing breeds.

Spring or Mid-season Lambing

This is the traditional system which matches lamb growth with grass growth and suits an all-grass farm. Requires good grassland management to ensure there is grass for tupping, grass for lambing and lactation, and grass for finishing lambs in summer. Unfinished lambs can be sold as stores or held over and sold in the spring as hoggets. Where there is plenty of grass it suits organic and low-input systems. Usually gets good lambing percentages and suits most breeds especially large crossbreds.

Late Lambing

Lambing in early summer (May) on grass produces low-cost quality fresh lamb during the following winter for a rising market. Late pregnancy and lambing are at a pleasant time of the year and no housing is necessary. Feed costs are low because grass is available before and after lambing. Can achieve good lambing percentages and suits most breeds. Possible problems include prolapsing and becoming cast (getting stuck on their backs) – both due to getting too fat on grass. Foxes can also be a problem. Lambs can finish on grass or be sold as stores.

Frequent Lambing

The flock is lambed every eight months – typically, January, September and May. In the UK the system is usually based on the Poll Dorset because of its long breeding season but other breeds can be induced to mate out of season with hormonal treatment (see Chapter 4). It requires very tight management, synchronized mating and reliable supplies of feed and labour. Small flocks can increase annual production without keeping more sheep, and produce prime lamb throughout the year.

may simply be an exercise in buying and selling but on the plus side it gives hands-on experience in stock rearing, handling and marketing. The ewes could be retained for breeding.

Organic Lamb

Most lamb is only a step away from being organic. The regulations allowing it to be labelled organic are based on normal welfare considerations plus restrictions on feed type, fertilizers and medicines. Management is difficult for small flocks unless they are part of a mixed farming enterprise.

Organic lamb is compatible with late lambing (*see* box, opposite) because late lambing flocks have a low demand for bought-in feed. Alternatively, many hill-bred lambs are naturally organic and could be identified and finished on organic lowland farms. The downside is that although inputs are low, so too are outputs because of low stocking rates; and not all abattoirs are registered to slaughter organic lamb.

A useful halfway stage – especially for local markets – is to call it 'naturally-reared'. Never call it 'hormone free' because all animals have hormones.

Sheepdog Training

Small flocks are sometimes kept for training sheepdogs. About twenty-five to thirty is the minimum size of flock needed and these, primarily for welfare reasons, should be dry sheep. One approach is to buy a flock of ewe lambs, keep them growing on grass, sell them for breeding a year later and then buy a new flock.

MANAGEMENT

Sheep are adaptable and can be bred, reared and sold at almost any time of the year; this means that a flock can virtually be established on any farm for any purpose. Normally flocks are managed on an annual cycle with the same activities at the same time each year. When setting up a flock it is essential to establish an initial plan – even though this will be revised and adjusted in subsequent years.

First, draw up a simple annual cycle (*see* Table 1, below) based on the following:

- The supplies of grassland (or other natural feed sources).
- When the products are to be sold.

Table 1 Annual Cycle of Sheep Production

Months	Activity	Feed	Labour
1	Rams in with the ewes	Moderate demand	Minimum demand
2,3,4,	Ewes in lamb	Moderate demand	Minimum demand
5	Ewes in late pregnancy	Increasing demand	Increasing demand
6	Lambing	High demand	Peak demand
7	Ewes rearing young lambs	Very high demand	High demand
8,9	Ewes rearing lambs	High demand	Moderate demand
10,11	Wean/sell lambs	Moderate demand	Moderate demand
12	Prepare for breeding	Moderate demand	Moderate demand

This table is a guide to the month-by-month activity plus the feed and labour demands in a flock. Details of feed requirements are in Chapter 5.

- A feasible lambing date to meet this market (*see* box, p.12).
- Other demands on labour and resources.

Flock Size

The number of sheep to keep is influenced by the amount of land available and how productive it is. The average number of adult sheep supported by a hectare (2.5 acres) of land for one year is called the stocking rate. For example, ten hectares of grassland supporting and rearing thirty ewes and their lambs for a year has a stocking rate of three ewes per hectare.

A stocking rate of twelve to fourteen ewes per hectare is not uncommon but keeping four or five ewes per hectare is a safe starting point on average lowland pasture. It may not sound many but there will be little grass during the winter and there will be lambs – perhaps two from each ewe – demanding grass in the summer. In upland conditions it could be four or five hectares per ewe.

Start with 20 per cent less than seems feasible. Disease builds up on heavily stocked grassland and the ideal is to have enough land to be able to rotate the flock and 'rest' their grazing areas every two to three years by grazing it with cattle, cropping it or cutting it for hay or silage.

Breeds

When choosing a breed the considerations for the small flock are:

- The shepherd likes the breed.
- It suits the environment.
- It suits the system.
- It provides an acceptable end-product.
- Docility.
- Cost and availability.
- Size. Small sheep are light to handle. Can keep more per hectare.

There are at least eighty-five pure breeds and recognized crosses in the UK alone and most suit a range of systems and environments. And there is often as much variation within as between breeds.

A ewe of any breed or cross will produce a slaughter lamb especially if it is bred to a Down ram (Fig 3). A safe bet is to choose a local breed or one from a similar environment. For other purposes such as wool production, rare breeds, pedigree flocks or dairying the choice of ewe breed is more specific.

Rams are basically of two types – terminal sires for producing a prime slaughter lamb, and crossing sires for producing breeding ewes. Terminal sires are well-muscled with a fast growth rate to produce fast-growing meaty lambs. Crossing sires will impart milkiness, prolificacy, hardiness and a good fleece in their female offspring. However, many breeds are dual purpose, capable of producing good breeding stock as well as slaughter lambs.

Commercial flocks producing prime lambs and breeding ewes will normally put rams of a different breed over the ewes. Rare breeds, pedigree, wool and dairy flocks will normally select rams of the same breed as the ewes.

Small, quick-maturing breeds produce lambs which reach slaughter weight quickly (ten to fourteen weeks) and are suited to early lambing flocks. Large, slow-maturing breeds can be used in later lambing flocks where lambs do not need to grow quickly to catch the early market.

BUYING SHEEP

The golden rules when buying sheep are:

- Buy from a known source.
- Make sure they are healthy.
- Get as much information about the flock as possible.
- Examine well before purchase.
- Buy ewes of similar type and age.

There are a number of sources of sheep.

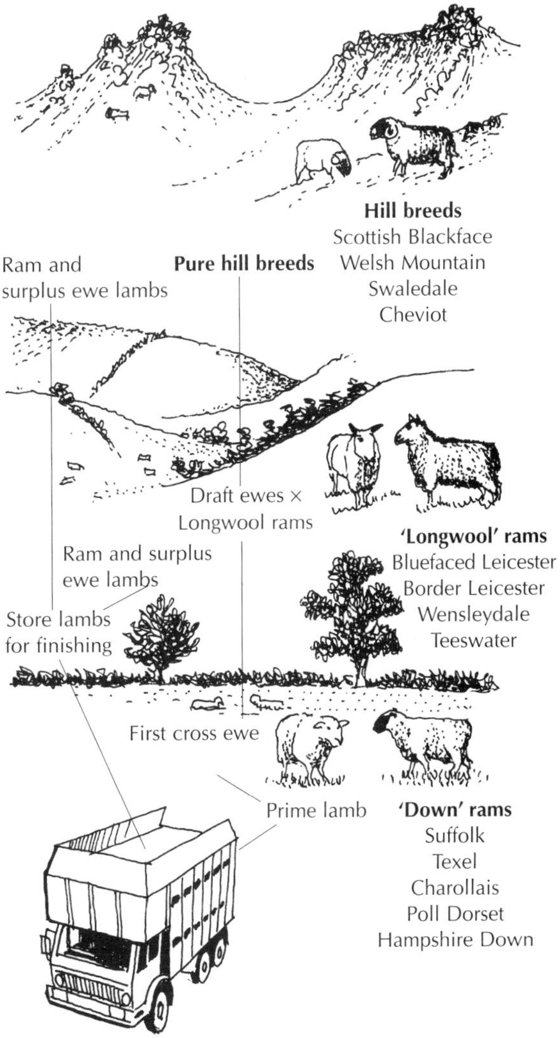

Fig 3 This illustrates the basic stratification of the UK sheep industry with a few examples of the numerous breeds and crosses. New breeds and breeding schemes have dulled the distinction between types of breeds, and Longwool rams (crossing sires) can sire prime slaughter lambs while Down rams (terminal sires) produce ewe lambs suitable for commercial breeding. Prime lambs will also come from dairy, specialist wool and rare breed flocks.

Some examples of 'first cross' ewes are:

North of England Mule (Bluefaced Leicester × Swaledale)
Scotch Halfbred (Border Leicester × North Country Cheviot)
Welsh Halfbred (Border Leicester × Welsh Mountain)

Hill breeds
Scottish Blackface
Welsh Mountain
Swaledale
Cheviot

Ram and surplus ewe lambs

Pure hill breeds

Draft ewes × Longwool rams

'Longwool' rams
Bluefaced Leicester
Border Leicester
Wensleydale
Teeswater

Ram and surplus ewe lambs

Store lambs for finishing

First cross ewe

Prime lamb

'Down' rams
Suffolk
Texel
Charollais
Poll Dorset
Hampshire Down

The important ones are:

- Direct from an established flock.
- From a market or specialist sales – usually in the autumn.
- Through a dealer/buyer who will source and transport them either direct from a farm or from a sale. Sheep contractors may buy stock.

Buying Direct

Buying direct from a farm is best because the buyer sees the sheep in their environment and can discuss their management and health care. The larger the flock the better the choice. Some flocks have annual reduction sales when they auction their surplus breeding stock. Look for advertisements in the local papers.

Breed societies have the names of pedigree breeders with stock for sale. When buying quality sheep – especially pedigree or dairy sheep – the records of the ewe should be available. These might record the amount of milk each dairy ewe gives in a lactation, how many lambs a ewe has each year or how fast the lambs grow.

Pedigree stock may have a fixed price according to age. Commercial ewe prices are often published in the farming press, or the local market auctioneers can advise. A factor in determining price is the number of lambs the ewe will have. A healthy one will produce five to seven lamb crops on average. So younger ewes cost more but will give more back. Ewe lambs tend to cost less because they are unproven and still have some growing to do.

Markets and Sales

Local papers and national farming publications will advertise markets in late summer when the major breeding sheep sales are looming. These auctions help to establish prices.

Stock sold at specialist sheep sales are often inspected by veterinary surgeons and come with some assurances that they are capable of breeding and have no important defects. Identifying the farm from which they come is useful as it may operate a health scheme and offer sheep which have been accredited (free from certain diseases).

Stock are penned in groups matched for size, appearance or age and are normally sold as a group (Fig 4).

Dealers

Auctioneers may recommend a dealer to select, bid for and transport sheep on behalf of a buyer. Some may advertise in the farming press or will be found at the markets. The dealer needs to be clear about what the buyer wants and how much he is prepared to pay. A good dealer is worth knowing, but a bad dealer can bring disease and disaster to the unwary.

WHAT TO BUY

Draft ewes

Many flocks are established from draft ewes. These are older, surplus ewes that are still capable of breeding. They may have come from hill flocks where conditions become too hard for them but they will thrive on lowland farms.

Draft ewes are experienced lambers, cheap producers of homebred stock and will eventually have a carcass value – although changes in slaughter regulations because of BSE may affect this. They may be at least a full mouth (*see* Appendix I, Fig 111) and possibly older. Draft hill ewes tend to be genuine. Check the reason for the sale of surplus ewes from lowland flocks and remember that few vendors will sell their best stock. Physical defects are obvious but rejects from a health scheme, barreners

Fig 4 Specialist markets sell breeding sheep in matched lots. Here, Polled Dorset ewes come under the hammer.

(who have failed to have a lamb) and problem lambers are not.

Young Ewes

Ewe lambs are less than a year old but may be ready for breeding. They are the cheapest of the young ewes, cost more in feeding because they still have some growing to do, and may not be easy to manage at lambing. An inexperienced shepherd may not want inexperienced ewes.

The next category of young ewe is the two-tooth (*see* Appendix I, Fig 111) They cost more than the ewe lamb, especially if they are proven breeders and are well grown.

Pregnant Ewes

Ewes are sold 'in-lamb' or described as having 'run with the ram'. In-lamb ewes should have been identified as pregnant by a pregnancy scan. Those running with the ram are not guaranteed as in-lamb. Pregnant ewes make a quick introduction to sheep keeping but it is vital to have accurate lambing dates otherwise their management both before and at lambing will be difficult.

Ewes and Lambs

Ewes with lambs at foot (sometimes described as couples) are often sold at local markets. The ewe is probably useless and the lamb it is suckling may not be her own. But they give instant experience into sheep rearing, handling and marketing.

Cade Lambs

Cade lambs (often called orphans) are surplus lambs from flocks whose ewes produce more than they can rear. Cades for sale should have had adequate colostrum (*see* Chapter 9), dry navels and be a few days old. They are sold at livestock markets but buying direct from a large flock is better.

They are often available quite cheaply to avoid the bother of having to hand rear them, which is labour intensive. Rearing these lambs will cost time, milk powder and lamb feed but they give experience in rearing and marketing or will make a tame breeding flock.

Rams

Rams, in breeding terms, are half the flock so it is important to get some guarantee or reassurance that they are fertile. Some vendors or sale organizers arrange fertility checks. Beware if buying an individual ram in a local livestock market that he is not vasectomized.

Sources of rams are a problem for the small flock. For getting started a cheap option is a good-looking uncastrated slaughter lamb out of a commercial flock. He will have hybrid vigour and should cost no more than slaughter lamb prices. Another option is an older ram from a commercial flock where sound rams may be routinely culled when they reach a certain age. Other options are discussed in Chapter 4.

Top rams from breeders may be performance tested – weighed and measured to ensure they have fast growth rate and lean meat. A small flock would need to share or be a serious breeder to justify the cost of one of this calibre.

Select a ram which is docile and easy to handle.

SELECTING SHEEP

Judge sheep on:

- Health.
- Physical appearance.

Usually, sheep look either healthy or dead; judging which is which is not that difficult. Commonsense says that a healthy sheep does not normally limp, hang its head, have droopy ears or dull eyes and does not have abnormal discharges from orifices.

Those which are carriers of infectious or chronic diseases are not obvious, so buying from flocks which are involved in health schemes is a reasonable safeguard.

The main physical health checks are on teeth, feet, udder or testicles and penis. They can be done with the animal on its feet, but for close inspection sit them up.

Checking the Ewe

1. The udder (*see* Appendix I, Fig 116) should be neat, soft and pliable, showing no signs of hardness or lumpiness. A dry ewe which has suffered mastitis may have lost the ability to produce milk but may not show signs of the disease. Feel carefully for a tell-tale lump at the base of the udder just above the teat.
2. Teats should be neat without cuts, warts or hard scar tissue. In maiden ewes and lambs the teats should be a sensible size. Very small teats – sometimes found on lambs from multiple births – can indicate a non-breeder.

   Four teats are not unusual but two of them will be supplementary and small. They frequently discharge milk but are rarely sucked. Alexander Graham Bell (of telephone fame) bred four-teated sheep to rear litters of more than two lambs, and many breeders have continued his work.

Checking the Ram

1. There should be two testicles (*see* Appendix I, Fig 115), both descended and of roughly equal size. They should be firm and slip up and down freely in the scrotum where there should be no swellings, cuts or lesions. The larger the circumference of the scrotum the greater the capacity of the testicles to produce sperm (Fig 5). The average circumference at the widest point in adult Down and Long-

wool breeds just before mating is 36–38cm. The average for ram lambs at eight months is 30cm.
2. Push the testicles to the bottom of the scrotum and feel the tail of the epididymis at the base of each one. This is where sperm is stored and should be firm and the size of a walnut.
3. The sheath around the penis should have no lesions or ulcerations and the penis should move freely within the prepuce (sheath). Check that the vermiform appendage (worm) is intact.
4. Semen tests are available through veterinary surgeons but they only show the fertility of the ram at the time of the test.

Teeth

Inspecting the teeth is important (Fig 6). Incisor teeth (*see* Appendix I, Figs 110 and 111) should be correct for the age, soundly in their sockets and aligned with the gums. Incisors should meet the pad within 5mm of the front edge. In sheep not expected to

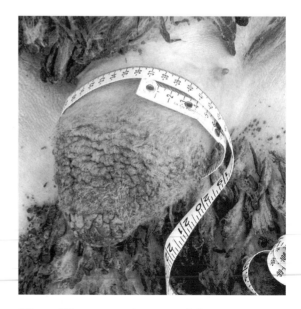

Fig 5 The circumference of the scrotum is a good indication of the fertility of a ram. Measure it at the widest part.

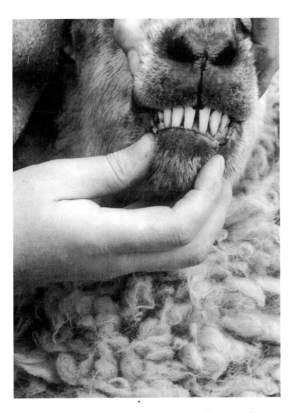

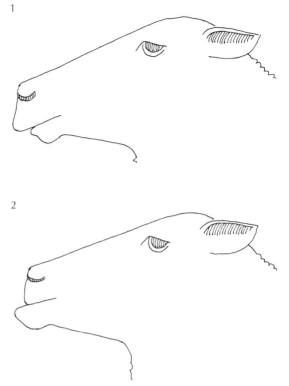

Fig 6 Checking the teeth. A full-mouth ewe – long in the tooth, well worn, but sound in the gums.

Fig 7 An undershot jaw (1). The front teeth do not line up with the pad in the upper jaw and eating may be difficult. Also avoid the opposite – the overshot jaw (2) where the lower jaw extends beyond the upper jaw.

eat root crops or graze too close, missing teeth (broken mouth) need not be a problem; although broken or pointed teeth can cause pain. Ewes which have lost all their incisor teeth (gummers) can do well on grass and could be a bargain if they are otherwise good sheep.

Damaged or loose molars (*see* Appendix I, Fig 111) mean ewes cannot cud properly and derive less from their food. A green stain around the lips is a tell-tale sign, otherwise feel their condition through the cheeks. Upper molars normally overlap the lower jaw and the overlap should feel smooth and curved.

Deformed jaws – overshot or undershot (Fig 7) – should be avoided especially if replacement ewes are to be bred from them.

Feet

Never buy sheep with bad feet. It is buying trouble. Foot care is hard work, sheep with problem feet suffer pain and the problem may be hereditary. They should look neat and not be overgrown or down on the pastern (Fig 8). Turn up each sheep and check feet individually for signs of separation, soft or pussed tissue, distortion or overgrowth (*see* Chapter 8).

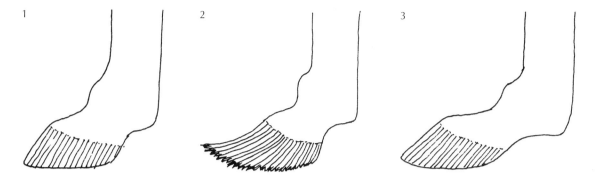

Fig 8 Start off with neat, trim hoofs and straight legs (1). Avoid overgrown hoofs (2) and weak pasterns (3).

Physical Appearance

Pedigree sheep will have appropriate breed characteristics, such as face colour, and the relevant breed societies will advise new buyers what to look for. Otherwise appearances are relatively unimportant. Some crossbred ewes may look odd but they have the advantage of hybrid vigour.

Do not fear thin sheep; unless they have problem teeth or diarrhoea they have probably given everything to their lambs. Equally, never be wooed by fat sheep. They may be poor milkers or even barreners.

BRINGING THE SHEEP HOME

Do not bring sheep onto the property until you have the basic facilities in place. These include:

- Insurance – primarily third party against straying, trespass and accidents.
- Adequate grazing.
- One bale of hay per head.
- Stock-proof fencing in all fields.
- Clean water supply.
- Sound gates.
- Shelter from the elements.
- A holding pen for the whole flock.

- Basic foot trimming equipment.
- Dosing gun and wormer.
- An airy building and a few bales of straw for a sick bay.
- A good torch.
- Waterproof clothing.
- A veterinary surgeon.

The golden rules when they arrive are:

- Isolate them from other sheep or stock for as long as possible.
- Give them a worm drench (*see* Chapter 8).
- Dip or inject them against external parasites if possible. Otherwise be vigilant about sheep scab.
- Have the veterinary surgeon check them. This also gives him the opportunity to see the stock and the set-up.
- Check their feet and either walk them through a footbath or spray their feet with a foot spray before putting them on grazing land.
- If appropriate, give any vaccinations that the previous owner may have used. Delay this for a week to reduce stress.
- Keep a close eye on them for a few days until they have settled.
- Make sure electric fencing is working well – especially if the stock are not used to it.

2 Fencing, Shelter and Housing

Fencing has two main purposes:

- To keep sheep on the property.
- To control their grazing within the property.

Sheep-proof fencing must have priority when setting up a flock. Few things cause flock owners more aggravation than straying sheep. They are dangerous, a waste of time, an embarrassment and could land the owner in court. Should sheep stray into other flocks there is the risk of spreading disease plus the problem of mobbing up and recovering them. On roads the owner is liable for any death or damage they cause.

Once sheep have found weaknesses in hedges and fences they rarely forget them. Gaps cannot be hidden by stuffing bedsteads or old netting in them. Once switched on to escape mode (usually activated by boredom or shortage of feed and more often in spring and autumn than in the heat of the summer) only good fencing will stop them. Bodged gaps can be dangerous – lambs can hang in loose netting and wire.

When sheep have strayed they like to return to their field by the route they used to escape – because this is what they remember – not by the route that the owner wants them to use. Always work with sheep and not against them.

Stone walls and hedges may look stock-proof but sheep will graze them, worry the parts where they can get a foothold and eventually create gaps. So fencing is often used alongside hedges and walls to protect their integrity.

Because hedges and walls also provide shelter, the fence can be erected close to a well-trimmed hedge which can then be allowed to grow through the fence. But beware that sheep can damage the fence when they feed on the vegetation. If trimmed hedges are preferred, the fence should be built one metre away to give a mechanical trimmer room to trim behind it, but on small properties and those with small fields this wastes grazing area.

The two main types of fencing used to control sheep are galvanized netting and electrified wire.

Barbed wire should not be used for sheep – they get tangled and injured but rarely restrained, especially when they are frightened. However, strands can top a fence to heighten it against cattle or horses and also protect against humans climbing over. Barbed wire wrapped around the top bar of a gate serves the same purpose.

Stock Netting

Post and galvanized netting (Fig 9) is a permanent system which, if erected properly using quality materials, will give at least twenty years' service with minimum maintenance. The initial cost is high but running costs are low. It is suitable for all fencing situations, especially boundaries. Lightweight, cheaper versions can be used in semi-permanent situations. Choose the correct version for sheep and lambs because the mesh is designed to prevent them pushing their heads through and getting stuck.

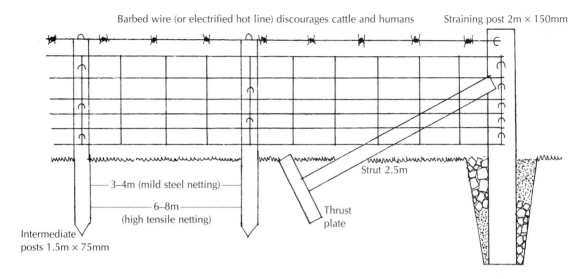

Barbed wire (or electrified hot line) discourages cattle and humans — Straining post 2m × 150mm

Strut 2.5m

3–4m (mild steel netting)

6–8m
(high tensile netting)

Thrust
plate

Intermediate
posts 1.5m × 75mm

Fig 9 Permanent stock netting fence. The illustration shows the basic construction and materials. From this it should be possible to work out quantities and costs. Advice on building the fence is usually available from the manufacturers of fencing wire, but the following is the basic sequence.

1. Set straining and any turning (change of direction) posts.
2. Set struts. Use as long a strut as possible and fix it halfway up the fence. A short, steep strut can lift the post out of the ground.
3. Use the top wire or netting rolled out and secured tightly at each end to give the line.
4. Set intermediate posts.
5. If using a top wire fix it in place.
6. Secure netting to end posts.
7. Strain netting.
8. Staple netting at every other wire from the bottom.
Straining and turning posts are set in a hole to lean slightly away from the strain, and earth and stones tamped in firmly. Intermediate posts can be hammered into holes made by a crowbar. (Not to scale.)

Sheep were badly designed; their ears fold back and allow the head through tight spaces, but prevent it from being withdrawn. Netting with oblong rather than square mesh is safer and note that in most designs the mesh gets smaller towards the bottom. Plenty of netting has been erected upside down; although on shooting estates it is sometimes done deliberately to give the pheasants an easy passage.

Wire netting (Fig 10) is a lightweight and cheaper version of stock netting and is ver-satile. It is used as permanent or tempo-rary fencing for boundaries, strip grazing, subdividing pastures for rotational graz-ing, rented grazing or temporary handling pens. It can be made vermin-proof by bury-ing the bottom edge.

Electric Fencing

Electric fencing (Fig 11) is versatile and lightweight and controls all types of stock on all terrains. It is used as a permanent,

Fig 10 The wire netting fence. High tensile wire can support the netting and allow wider post spaces. The bottom can be buried to prevent predators from getting underneath. (Not to scale.)

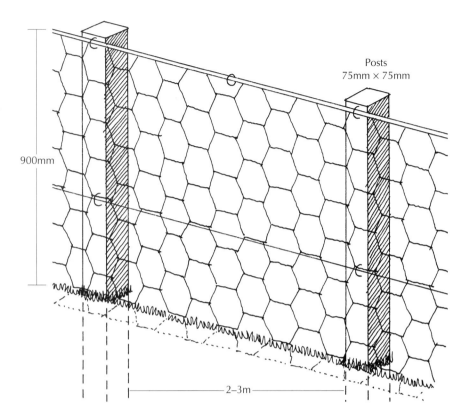

Posts
75mm × 75mm

900mm

2–3m

Fig 11 How the electric fence works. When the ewe touches the fence – especially with her nose or ears – she completes the electric circuit and gets a short shock.

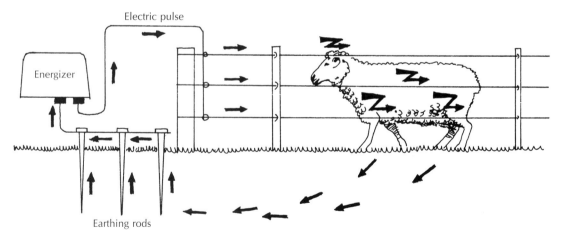

Electric pulse

Energizer

Earthing rods

temporary or portable system and is ideal for fencing rented grazing. The initial outlay can be half that of post and netting fence and, set up correctly and using good materials, it can last for at least ten years. Permanent systems use insulated timber posts whilst the temporary and portable systems use plastic or insulated metal posts.

Electrified systems are attractive to the new flock owner who does not want to commit too much money to the project; redundant equipment retains a good secondhand

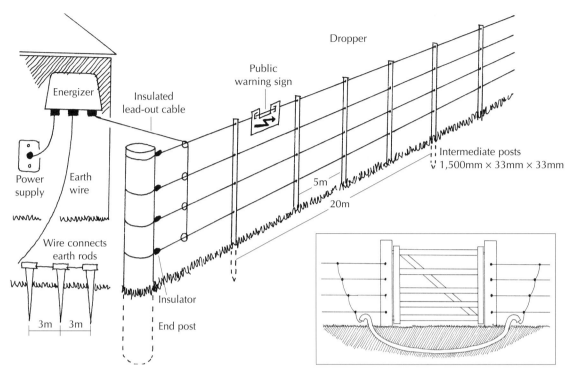

Fig 12 The permanent electric fencing system. Use high tensile 2.5mm galvanized wire. Electric fencing need not be tightly strained so end posts do not have to be strutted. Wood posts and droppers need insulators unless they are special self-insulated material in which case the wires can go through them. Earth rods should be 2m deep for mains electricity systems. Droppers keep the lines evenly spaced and just rest on the ground. End posts are set in a dug hole then earth and stones tamped firmly around them. Intermediate posts have a sharpened end and are sledge-hammered into a small pilot hole made with a crowbar.

The inset shows how to take power across gateways and troughs. Bury polythene water pipe about 0.3m underground to carry the under-gate cable which is available from electric fence manufacturers. Turn down the ends of the pipe to prevent it from filling with rainwater. (Not to scale.)

value. Because sheep are naturally well insulated by their wool it is essential to invest in good equipment.

Permanent systems (Fig 12) are usually run from the mains power supplies but temporary fences (Fig 13) and those on off-lying land can be run from batteries (wet cell and dry cell) and wind or solar panel chargers are available. Old car batteries are suitable for wet cell energizers but always have two available – one working and the other charg-

ing. Try to avoid dry batteries because they are an expensive source of power.

Modern energizers can belt out 5,000 volts and power 100km of fence line but are not hazardous because the electricity is pulsed to give a short sharp shock from which the victim recovers. The pulses last for 350 millionths of a second and are spaced about one second apart. Never incorporate barbed wire in an electric fence and never power one fence with two fencers.

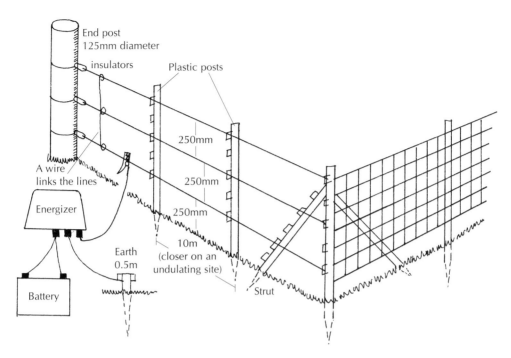

*Fig 13 A portable or temporary electric fencing system. This illustrates two systems –
polythene wire and polythene netting. Electric netting is usually supplied with its own
posts. Plastic posts for the wire system have multiple lugs to hold the wire at variable
heights. Metal posts have adjustable insulators. Having the bottom wire too close to the
ground can lower the voltage. Polythene wire has a high resistance to current flow and
should only be used for fence lengths of up to 1km. Use seven strand galvanized wire for
longer temporary fences. Wires should be linked electrically every 400m.*

Keeping the power on permanently – even
when there are no sheep in the field – burns
back invasive vegetation and deters rodents
from eating any plastic components.

When sheep have been trained to respect
electric fencing, just a single line at nose
level, alongside a wall or hedge, can suffice.
Sheep are led by their noses and damp
noses are the most vulnerable to shock. A
flock which is unused to electric fencing can
be trained by putting a temporary fence
300mm from a secure fence or hedge and
putting feed close to it to lure them to touch
it. Homebred lambs learn from birth.

Stone walls can be sheep-proofed with a
single 'hot line' projected about 300mm
from the wall (Fig 14).

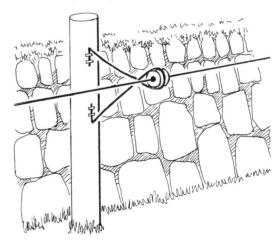

*Fig 14 The hot line is a simple way to
reinforce walls, hedges and fences. The
line should be supported about 300mm
from the wall.*

25

Although it may deter foxes, electrified fencing around lambing paddocks is not a good idea. Lambing ewes like the privacy of field perimeters and, in the oblivion of birth, can lamb up against them or have the new-born lamb get to the wrong side.

Electric fencing is not a good physical barrier so, where fields are subdivided or strip grazed, electrified poly netting (Fig 13) is often used because it is a visual barrier. It also has the advantage of being effective against most predators and rabbits but must not be used with horned sheep.

Building a Fence

Fencing is a skill and the stock netting system that requires considerable straining, or the permanent electric fencing that needs to be reliable and secure, should be erected by a skilled fencer. Fencing contractors have the necessary tools, experience and physical fitness.

However, temporary electric fencing and wire netting systems are excellent training grounds for the first-time fencer. Their relative cheapness, versatility and effectiveness mean that the flock owner can learn about fencing whilst keeping the stock under control.

Temporary fencing is useful to subdivide a large field to make grazing and stock handling easier. Changes can be made to the layout and, when perfected, can be replaced with permanent fencing.

The fencing illustrations in this chapter show only the basic components and construction of a range of sheep fencing systems; manufacturers of fencing equipment usually publish booklets describing fence building in detail, and materials and advice are available from local agricultural and specialist fencing merchants.

There are variations in the quality of products – some timber posts are better preserved than others and some wire is better galvanized, so check for the correct

British Standard or equivalent. In general terms the costs of erecting a fence are around one-third for wire, one-third for posts and one-third for labour; so quality in all departments makes sense.

The steps to fencing are:

- Decide what types of fencing are needed.
- Take into account any other species they may have to control because this can affect design and construction.
- Pace out the fence lines (straight fences use less materials).
- Check that boundary lines are correct to avoid encroaching on neighbouring land.
- Draw a sketch map of the fence lines showing gateways and water troughs and where the posts and other components will go. Also take account of public rights of way such as footpaths and stiles.
- Calculate how much of each component is needed.

Maintenance

Fence maintenance is an important part of the shepherd's routine. Fence lines should be checked regularly before the sheep discover any weaknesses. Replacing the odd post or length of wire is a good way to learn the art of fencing.

Look for signs of rust or broken strands in the wires (especially where they attach to the posts) and rotting or loose posts. Staples are used widely in fences and fence repairs. Use galvanized ones of the correct size, and angle them so that both points do not go into the same grain of timber and split it. Do not drive them in tight because this breaks the galvanized coating on the wire and accelerates rusting. Fence repair may also involve connecting broken wires for which proprietary connectors are available.

Checking electric fences each day is a wise habit. Sheep appear to be able to sniff when the power is off without actually touching the wire.

Where an electric fence is not giving a shock there are several fundamental reasons:

- The power is not connected or the battery is flat.
- A wire has broken (therefore breaking the circuit). Metal strands in polythene wire may be broken but the plastic is still intact. This is not easy to see. Strands often burn at joins, which may be visibly scorched.
- A wire has shorted out on some metal. Often found where electric fencing is installed alongside old wire fences. And during gales, galvanized roof sheets may blow onto the fence. Heavy vegetation, especially in wet conditions, may short it out and flatten the battery. Some users like to trim or spray under the bottom line.
- The system is insufficiently earthed. This is a common problem.
- The insulators have broken down.
- There is a poor connection somewhere in the system.

Fence testers are available and those with a digital read-out are recommended. Alternatively, push a metal fencing stake into the ground within 5cm of the fence wire, hold it by an insulator and push the stake against the wire. The length and sound of the spark that jumps to the earthed stake gives an idea of its efficiency. To test if the fence is working, touch it with a blade of grass; this reduces the shock to a comfortable level.

Safety

There is also a safety element to fencing. Dangerous areas such as sheep dips should be fenced off, especially against children and trespassers. Chain link fence is recommended as the least climbable, and if less than 2m high should be topped by two strands of barbed wire. Electric fences, especially where there are public rights of way, should have warning signs attached. Fencing companies should be able to give up to date advice on the law.

Stockproof Hedges

Hawthorn, and to a slightly lesser extent blackthorn, are most suited to stockproof hedges. Both have thorns to deter browsing sheep but the blackthorn puts out invasive suckers so a mix of one blackthorn to every two or three hawthorn is usual. For variety some holly, willow, oak and crab-apple can be added. Such hedges need to be well maintained after establishment to prevent gaps at the base. New hawthorn and blackthorn hedges respond to regular cutting with a flail mower and an annual trim along the sides and the top produces fresh bushy growth.

Unless reinforced by fencing, hedges are probably best used around internal fields where strays may cause less problems, rather than as farm boundaries. Cultivate and fallow the site to be planted for one season, and plant in November through a black polythene mulch strip to control weeds and aid moisture.

SHELTER

Sheep must always have access to shelter from the elements. Lack of shelter increases their demand for food, and shelter is vital at lambing time. It is also important at tupping (mating) time when extremes of weather can reduce conception rates.

Sheep unable to find shelter from the sun often stand in a huddle shading their heads under each other's flanks or, typically, they stand in a circle like the spokes of a bicycle wheel with their heads at the centre in the shade. Apart from distress and reduced grazing, excessive heat can cause hyperthermia and will reduce fertility, especially in rams.

Shelter from wind, rain, snow and heat may be provided by a belt of trees or shrubs, a hedge or wall, or artificial wind-break material.

Shelters must be secure otherwise they will cause more harm than they seek to prevent. Equally, sheltered areas can result in a build-up of mud and disease.

Where there is no natural shelter, artificial windbreaks can be constructed from plastic windbreak material which has 50 per cent permeability and is designed for livestock (Fig 15). Heavy-duty stockproof windbreak material can keep wind speed down to below 7mph. On the sheep farm a roll of this material has many uses, serving as temporary penning, ventilation in buildings and shelter belts in fields.

The main purpose of a windbreak is to reduce the wind speed rather than stop it. Useful shelter is found for a distance of twenty times the height of the barrier downwind, but often sheep stand happily on the upwind side of a hedge. This is because the wind strength reduces at their level as it lifts to go over the hedge – giving some shelter for about three times the height of a barrier.

Woods or shrub shelter belts should be reasonably dense otherwise they are draughty at the base. Plant them across the prevailing wind – or the wind direction for a crucial time of year such as lambing – and just over the brow of an elevation. Use a variety of trees and shrubs with taller trees, broadleaves and conifers in the centre, and shrubs and smaller trees towards the edge to give shelter at ground level and an aerodynamic profile.

Choose species which grow well in the locality, get small specimens (large trees have lower survival rates) and plant from autumn to early spring. Coppice the small shrubs and trees to maintain dense cover.

HOUSING

Some form of basic housing is important if only to protect the flock and shepherd during routine work such as shearing and vaccination, and some housing space is vital for nursing sick animals. A sheep house merely protects against the elements and need not have a higher temperature than outside.

Where sheep are seriously housed the standard housing period is from eight to twelve weeks before lambing, until turnout

Fig 15 Light and easy to handle, plastic windbreak material has numerous uses for shelter, temporary penning, covering the open sides of buildings, and for attaching to the bottom of gates to stop lambs escaping.

with their lambs – which normally covers the winter and the lambing period. Temporary or short-term housing may simply house the flock during lambing and for two days afterwards.

The advantages of housing are:

- Increased stocking rate because the flock is not limited by winter grazing.
- Keeps the flock off the grass in the winter to give extra grass in the spring for ewes and lambs. Grass growth is delayed and reduced by 15–20 per cent if swards are grazed during January to March.
- Easier shepherding during the winter and at lambing.
- Cleaner grass because it has a break from winter worm burdens (see Chapter 8).
- Better performance because nutrition, health care and lambing are better supervised. There are claims of increases in lamb survival of 30 per cent.
- Less food waste.
- No losses from predators.
- Older ewes can be retained and flock replacement costs reduced.

The disadvantages are:

- Cost – such as the cost of the house, equipment, bedding and services.
- Risk of disease.
- Disposal of manure.
- More expensive maintenance feed in the form of hay or silage.

Costings usually show that housing is economically viable in large flocks, and in smaller flocks it can at least cover its own costs provided the building is kept low cost.

As a first step to housing, sheep can be wintered in open yards but must have shelter, excellent drainage and low rainfall. Temporary buildings constructed from big bales are another option but are better suited to arable areas where the bales are available at low cost. A compromise is a shed which opens onto fields from which the flock

has free access. The problem here is condensation from wet fleeces, and muddied access.

Stone buildings and lean-to barns will convert cheaply to sheep housing but attention must be given to hygiene and ventilation. Windbreak cladding can be used on almost any building to provide weatherproofing and good ventilation, and timber from secondhand wooden pallets make pen barriers and feed troughs (Figs 18 and 19).

When planning a building consider:

- The numbers to be housed.
- Feed system.
- Length of housing period.
- Type of floor and bedding.
- If planning permission is needed.
- Any other uses the building may have when not being used for sheep.
- If lambing indoors ensure there is enough space to hold ewes and lambs if the weather is too bad for turnout. Ewes with lambs need twice the space of unlambed ewes (see later).

The main considerations when building or adapting a sheep house are:

- Siting – preferably naturally sheltered; easy access; protected from rainwater running inside.
- Ample draught-free ventilation.
- Fixtures and fittings can be kept clean and cannot cause injury.
- Shape should allow suitable penning, working and storage area.
- Roof should have 3–4m clearance.
- Walls solid to at least 1.2m high.
- Floor porous, free-draining and dry. Concrete is suitable and easy to clean but will need extra bedding.
- Materials non-toxic.

A lean-to or mono-pitch pole barn (Fig 16) is a basic and versatile sheep house. Fronting onto a concreted yard it can be the centre of a sheep enterprise. The lower portion of the walls should be a solid construction of

Fig 16 A simple lean-to shed for winter housing or lambing with a plan of a suitable layout. Pens could be further subdivided by walk-through troughs if smaller groups of ewes were wanted. Using hurdles and basic feed barriers and troughs (Figs 18 and 19) the shed can be cleared and used for other purposes.

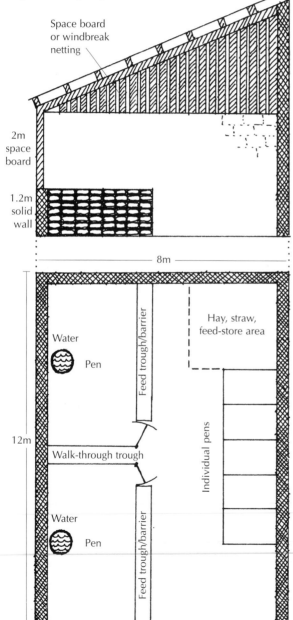

blocks or galvanized sheeting to at least 1–1.2m above the final level of the bedding. Eaves should be not less than 3–4m above ground with ventilation between the eaves and lower wall and at the gable ends. Ventilation may be provided by timber space boarding or plastic windbreak cladding, both of which reduce the draught and protect from rain and snow.

Open barns can be converted to sheep housing by constructing secure bale walls to 1–2m and topped with plastic windbreak cladding.

Polythene Tunnels

Polythene tunnels (Fig 17) are one of the cheapest forms of stock housing. Specialist stock tunnels that incorporate ventilation systems are necessary for large flocks but for the small flock a 10–12m horticultural tunnel – open at both ends for ventilation – is ideal for lambing and other short-term housing needs such as finishing lambs and shearing. Windbreak netting or wooden barriers inside protect the polythene from sheep, and sheet shades are available to reduce the temperature on sunny days. An earth floor is ideal provided it is sited on free-draining soil. Drainage can be assisted by having the floor higher at the centre and sloping to the sides. Sunlight sterilizes the soil and eliminates damp corners.

When not needed for sheep they provide a storage area (beware of condensation dripping off the roof), fertile soil for growing horticultural crops or space for rearing turkeys in the run-up to Christmas.

Tips for getting the best from a polytunnel are:

- Site it where it is sheltered from sun and wind, but not under trees because of the risk from falling branches.
- Set it up on a warm and windless day.
- Use insulating foam on the frames to protect the polythene against heat and protrusions.

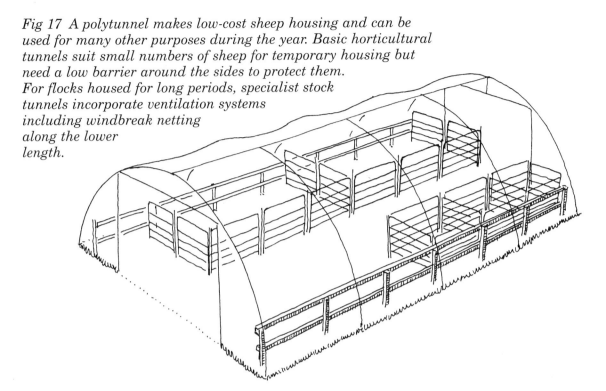

Fig 17 A polytunnel makes low-cost sheep housing and can be used for many other purposes during the year. Basic horticultural tunnels suit small numbers of sheep for temporary housing but need a low barrier around the sides to protect them. For flocks housed for long periods, specialist stock tunnels incorporate ventilation systems including windbreak netting along the lower length.

- Tension the polythene very hard to prevent chafing.
- Secure the frames and the edges of the polythene very firmly.
- Keep a roll of mending tape handy to cover small holes. Birds damage the polythene by pecking at insects that they see on the inside and cats use them for scratching posts.
- Do not wash off any greening on the roof; it extends the life of the polythene.
- Galvanized frames have a long life but, if they become corroded by dung, treat them with a bitumen paint.

Ventilation

Ventilation is the single most important consideration. A sheep house must not be stuffy, suffer condensation or smell of ammonia.

Ventilation should be at a high level and never create a draught – especially in the lambing area. Natural methods need an open ridge plus high level ventilation in all walls. The inlet ventilation at the eaves should be equal to twice the outlet ventilation at the apex. The simpler the building and the less internal obstructions the better the ventilation. The eaves should be at a minimum height of 3m.

Ventilation of old stone buildings may not be easy. An outlet at the apex is important, otherwise get sufficient ventilation through the four walls, and open the doors on calm days – preferably end to end.

Flooring

Concrete is an expensive surface for housed sheep, but a concreted working area and concrete floors for mothering-on pens help with hygiene. A concrete floor should have a 1:100 fall for drainage. Slatted floors are widely used for large flocks but are rarely relevant in small flocks where they can double the cost of a building and reduce its versatility.

31

Ideal floors are porous with 150mm of chalk, gravel, rubble, hardcore, round pebbles or stone (50mm to dust grade) but must not be sharp or awkward to walk on and should be on free-draining soil.

Straw is the best bedding for sheep who use around 1–1.5kg per head per day. On porous floors allow one bale of straw per ewe per six-week period – more if on concrete. Spread a third of the estimated needs at housing then top up daily. Sawdust and shavings are not suitable at lambing because ewes do not like licking lambs covered in sawdust. Peat or horticultural alternatives have been used successfully in small flocks.

Space

Housed sheep should have enough space to practise normal behaviour and to have exercise. The standard recommendation is that ewes should be housed in pens of twenty-five to forty with around 1.5–2sq m of floor space per ewe. Non-breeding sheep can be housed in larger groups. Store lambs need up to one square metre per head and six-week-old lambs about 0.5sq m.

Layout

Trough space usually dictates the shape and size of pens. Sheep are flock feeders and pens need to be long and narrow to allow enough trough space along one side for all the flock to feed at one time. Troughs on opposite sides can result in ewes dashing dangerously from one side to the other. Allowing 500mm per ewe requires a pen 12m × 3m for twenty-five ewes, but any other permutation to suit the shape of a building is possible provided the principles are met.

Pens can be constructed from wooden rails and be 1–1.2m high. Sheep and shepherds must be able to move in and out of the pens easily, so easy-to-open gates or hurdles in the corners are essential.

Putting the feed outside the pen so that sheep eat through the barriers is a simple system. The passage outside the pen needs to be at least 1.25m wide (excluding the

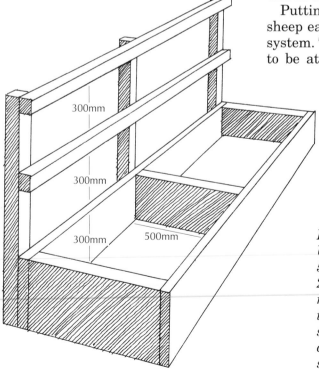

Fig 18 Wooden feed barrier. Uprights and rails are 50mm × 50mm timber and the planking is 25mm thick. Small breeds may need a slightly lower trough. The trough is not essential because sheep can feed through the barrier and direct from the floor. (Not to scale.)

32

trough) for hand feeding and using a barrow. Troughs and barriers (Fig 18) can be metal or wood and must not have rough edges. It is best to make them portable so that the building can convert to other uses and be easily cleaned.

Where smaller pens and groups are preferred the pen can be subdivided by walk-through feed troughs (Fig 19) which double as barriers and feeding troughs – especially for hay and silage – and an observation area.

When feeding hay, silage or straw *ad lib* allow 100mm space per head. Feed at as low a level as possible because sheep eat with their nose to the ground and have difficulty in swallowing when the head is up. When they pull a mouthful from an overhead rack they lower their head to eat it – wasting some of it and shaking debris and seeds onto the fleece and in the eyes, causing contamination and inflammation.

Other provisions include an area for individual mothering-on pens for newly-lambed ewes. The standard ratio is one pen per six to ten ewes (*see* Chapter 9). It saves labour if the building also stores bedding,

hay and other feeds but they must be protected from vermin and birds.

Water

Sheep must have twenty-four-hour access to clean fresh water and need at least five litres a day during lactation. They are fastidious drinkers and will stop drinking rather than drink dirty water.

Water can be supplied to the small flock in troughs or buckets which should be emptied regularly to prevent clean water being added to soiled water; white buckets show up any contamination. Standard bowl drinkers which are plumbed into the water supply are the easiest system but are not always popular with the sheep.

Important points for a water supply are:

• Troughs should be raised so that sheep have to stand to drink – preferably on a wood or concrete block platform – to avoid fouling. A lower supply is necessary for lambs but make sure they cannot fall in and drown.

Fig 19 The walk-through feed trough is a double-sided feeder for hay, silage and concentrates. It also subdivides pens and is an observation area. Uprights and rails are 50mm × 50mm timber and the planking is 25mm thick.

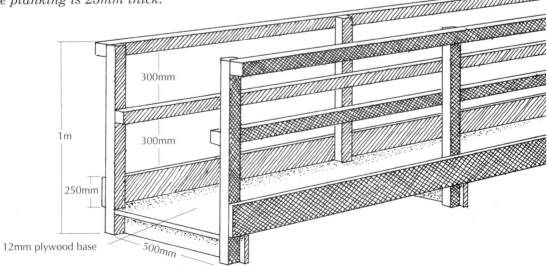

- Sheep prefer to drink from open troughs rather than bowls.
- Protect from freezing.
- Hill sheep in particular prefer running water which can be supplied through guttering along the length of the house.
- Troughs should be sited at the side or back of a pen rather than in a corner where bullies and other sheep can block it; and away from the risk of freezing.
- Containers must be easy to keep filled.
- Sheep are messy drinkers so the area must be kept dry to avoid foot problems.
- Water should be sited where it cannot be contaminated with feed.

Health

Under good stockmanship where stress is kept to a minimum, housed sheep suffer few health problems. Those which may arise are pneumonia, footrot and external parasites.

Pneumonia is triggered by poor ventilation and by stress, so as a rule:

- Do not house when the fleeces are wet.
- Do not feed or bed on dusty hay or straw.
- Avoid sudden change by introducing their 'housed' diet at least two weeks before housing.
- Inspect and treat the whole flock for footrot (*see* Chapter 8) before housing. Infection spreads in wet bedding and can cause deadly liver abscesses in young lambs.

Some flock owners give a worm drench at housing to reduce the worm burden and may control external parasites such as lice by dipping or pour-ons.

Behaviour

Most sheep adapt well to being confined although close-flocking breeds like the Romney may be happier than the more individualistic breeds like the Scottish Blackface.

Ewes which are bullied by dominant sheep cannot escape so avoid mixing different groups, ages and breeds where size and behaviour might vary, and avoid mixing polled and horned sheep.

Housed sheep should be able to lie or stand when they choose and in the orientation they choose (usually parallel to the pen sides) and drink and eat when they wish (usually when the rest of the flock want to drink and eat). They should not be subjected to extremes of temperature, noise, bullying, damp bedding or food and water deprivation.

Wool-biting is peculiar to housed sheep, although rare. Lower-ranked sheep are the victims and they cannot escape their biters. It is not stopped by the provision of fibrous food and it seems to induce higher stomach worm burdens in victims – probably involving stress-induced immunodeficiency.

Electricity

Housed sheep should be clearly seen and need a general lighting level of 50 lux. A few translucent roof sheets and open ends usually give enough daylight. Fluorescent lights are costly to install but cheaper to run than ordinary light bulbs. It is best to have a system whereby some lights can be left on at night so that the flock is not spooked when all the lights are suddenly switched on.

Sockets should be available for an infrared lamp in a mothering-on pen and for a kettle for hot water. Keep all wiring out of reach of sheep. Weather-proof electrical fittings are recommended for livestock houses because of the dampness in the air and the use of hoses when cleaning out.

Safety

Stock houses which combine straw, hay, electricity, infra red lamps and humans are dangerous. Make sure that electrical wiring is safe and that there is a quick exit for all animals in the event of a fire.

3 Moving, Handling and Transport

Moving, penning and handling sheep are the most physical and stressful jobs on a sheep farm. Good handling – which is essential if some husbandry jobs are not to be neglected – is the ability to control sheep and persuade them to do what you want them to do with minimum fuss.

It is vital that, once under control, neither the flock nor an individual is allowed to escape. Penned sheep that break out, or an individual that has escaped the clutches of the shepherd, will be trebly hard to get back under control.

The key is to:

- Work with the sheep – not against them.
- Understand their behaviour.
- Plan ahead.
- Have secure handling equipment.

A typical sheep handling sequence would be to mob them in a field; drive them to a handling area; pen, handle and work on them; load them into transport.

Mobbing

Sheep have a natural instinct to flock and follow, which is why one person can handle large numbers.

Sheep are alerted by what they can hear, but tend to respond to what they can see. Because sheep are preyed upon, they have developed panoramic vision – allowing them to see moving objects at a distance but not in detail; thus they respond to a moving object but not to a stationary one. As a result, sheep will rush together to form a flock when they are startled by noise or movement. The flock may then stand facing the 'danger' or, if the disturbance is friendly, such as a shepherd with food, will rush towards him.

A flock can be controlled by utilizing its flight zone – the personal space immediately around it – and will move away from anything that enters that space. The more tame an animal or flock, the smaller the flight zone. By finding this zone a shepherd can keep the balance between the flock fleeing and standing still, and can move them in the direction he wants. He can direct it with an arm or outstretched crook; a loud or abrupt sound will make them move but a softer, prolonged sound can immobilize them.

He can judge the risk of fleeing by the proportion of sheep looking at him. The more that face away, the closer they are to flight.

Driving

When a flock is mobbed and ready for driving check that none has been missed, either by counting or by checking the field perimeter.

Sheep on the move need something to aim for and provided they associate the shepherd with food and not fear they will move towards him.

Sheep prefer to follow rather than to be driven, so flocks are more easily moved if the shepherd walks ahead with a pocketful of sheep feed and accompanied by a friendly 'Judas' sheep.

Always work quietly, patiently and with vigilance so that any plan to break away can be anticipated and stopped. Sheep behaviour is predictable, but frightened sheep become unpredictable. Running sheep are usually out of control.

Sheep prefer to move uphill and towards the horizon and freedom. So gateways should be sited so that they are obvious and show open space beyond. Gateways sited in the corner of a field have the added advantage of funnelling the flock and making a useful temporary penning area (Fig 20).

Sheep are inquisitive and baulk at changes in the surface under foot. They are reluctant to go through a gateway where grass changes to mud, ruts or puddles. Sheep are nervous of water and lock their front legs stiff as an instinct against falling, which stops the flow of sheep. They creep around the edges of puddles or leap from hard patch to hard patch. Levelled-off hardcore in gateways is invaluable, and deep ruts and mud should be avoided.

When in doubt sheep try to go back the way they came, so as they go through a gate there is a risk that they will turn and run back. If a flock is being driven through a gateway, follow about three body lengths behind to allow enough distance to stop and turn them. Make sure that gates open back tight against a hedge or fence to prevent sheep getting behind them.

A route which takes them alongside hedges, fences or buildings helps to direct them. Take care going downhill, particularly if they are pregnant; most of their weight is over their rear legs and they find it less easy to go downhill than to go uphill.

Sheep are unhappy walking in strong winds and heavy rain and may seek shelter or even refuse to move.

Moving ewes and young lambs is an exercise in patience. The mob rotates and makes

Fig 20 *A temporary pen, utilizing a few hurdles, can be taken to the flock rather than taking the flock to the pen. A corner gateway makes a good site. On a frequently used site a few permanent posts add strength, and some posts and netting can be left permanently for a funnel. Once the sheep are confined the hurdles can be closed up behind them to form a catching pen. The sheep can exit between any hurdle. Spare hurdles can make a second pen.*

very slow forward progress as ewes and lambs try to keep in contact. A lamb, separated from its dam, instinctively runs back to the field where it last saw her. When this happens there is usually no alternative but to return the whole flock to the field and start again.

To avoid the problem, make sure that all the ewes and lambs are mobbed, walk them very slowly and do not follow too close behind so that the shepherd has time and room to react to a lamb turning back. At gateways and in lanes, follow with a length of netting between two people to block their escape. Make sure gates are low to the ground or have netting on them to prevent lambs getting back through after they close. A young lamb's homing instinct is remarkably strong and will persist for the whole journey.

Penning

Sheep get used to routine handling and if penning is not associated with pain or fear (for which they have very good memories) they can be penned quite easily. Again, a bucket of feed and a Judas sheep will encourage most of the flock in. Have a pen large enough to hold the whole flock easily – allow at least half a square metre per sheep.

A rectangular shaped pen is better than a square one because they flow to one end rather than around the sides and back out again; and they should be able to see through the sides so that they do not feel trapped. Again, they are wary of a change of surface, and do not like going from light to dark – such as into a building.

Do not let the flock into the pen until the sheep are well mobbed up, otherwise the first ones decide they do not like it and run out, turning the followers; they tend to stop 3–5m from a dead end and retrace their steps. Get them well mobbed and, provided they are not pregnant or have baby lambs, rush them at the last moment by shouting and waving. Used strategically, such as at

pens or gateways, noise and action work because sheep flee from noise.

Finally, have a gate which closes quickly and securely behind them.

Handling

Sheep are easier to catch and handle if they are closely penned. The traditional way is to fill a smaller pen – a catching pen – from the holding or gathering pen (Fig 21).

They need a reason to go into the catching pen; for example they are tricked into seeing an exit ahead of them or they follow a feed bucket. Decoy sheep and mirrors have been used as lures but both have the drawback that they face the oncoming sheep; this is seen by the sheep as aggression and will halt it. Normally they will only follow another sheep when they see its rump ahead of them.

Although sheep like to move in a straight line, their flow through a penning system can be maintained by having curves or slight corners which give the illusion of the sheep ahead disappearing out of sight.

If an individual sheep turns in a restricted area and faces against the flow there is normally nothing to be done except physically turn it to face the right way.

Once penned, sheep are handled as individuals or small groups and other behaviour patterns come into play.

Groups of sheep of less than four do not act as a flock and are difficult to control.

Individual sheep always try to maintain visual contact with at least one sheep. If they lose contact they try to restore it quickly – even to the extent of knocking down a person standing in the way.

Sheep innately fear isolation, and need companionship. Where an individual needs to be isolated (unless it is sick) it should have a companion. Fear, measured by levels of the appropriate hormones, has been found to be higher in sheep which are on their own than in sheep that are in sight of a slaughtering area.

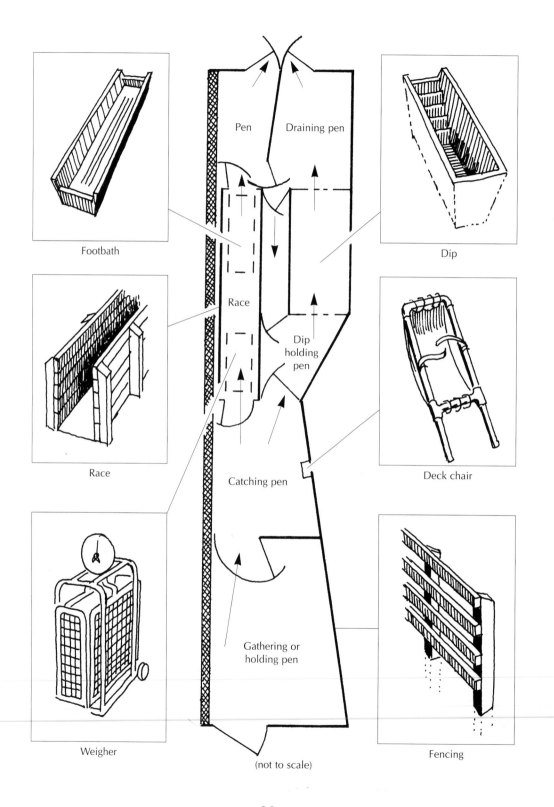

Footbath

Dip

Race

Deck chair

Weigher

Fencing

Pen

Draining pen

Race

Dip holding pen

Catching pen

Gathering or holding pen

(not to scale)

*Fig 21 The layout and flow for a basic permanent handling and treatment system.
The layout can follow a straight line or may go around the corner of a building to utilize
the walls and to add curves and bends which help the flow of sheep. The dip system can
be omitted. The flow (shown by the arrows) gives the facility for recycling sheep. Hurdles
may be used in the system, especially as gates.*

*The footbath can be used as a stand-in type or walk-through and may be permanent or
removable. The race should be 900mm high. Internal width is 0.5m. The sheep weigher is
removable. A small plunge-type dip will be adequate; they can be home built from
concrete blocks, or fibreglass ones are available which are sunk into the ground. Sheep
deck chairs are a simple system for restraining and working on sheep. For easier
handling a turnover crate may be added to the exit of the race.*

*Timber rails 75mm × 50mm and posts 1.5m × 100mm × 100mm set 1.5m apart and at
least 600mm in the ground make a strong permanent external fence. Four rails are
sufficient and at the bottom the gaps between them should not exceed 100–125mm to
prevent lambs escaping.*

Loading

Loading is the most stressful part of transporting sheep. It is against their nature to
be forced into a dark, confined space. The
ones to be loaded should be separated and
the rest of the flock taken away. Encourage
the group with food or with a decoy sheep
which must face away from the following
group. A view of freedom through the front
of the vehicle encourages them. The back of
the vehicle must close quickly and securely.

Low trailers may not need a loading
ramp if sheep jump on and off freely and
safely but it may be required by law if the
trailer is more than 310mm off the ground.
With high vehicles such as pick-up trucks
sheep can be lifted in and out, but a non-
slip loading ramp with a gradient of around
1:2 makes life easier.

Eliminate gaps between the ramp and
vehicle or between the lowered tailboard
and vehicle because they break legs.

Holding and Lifting

For most operations individual sheep are
restrained on their feet. A hand under the
chin is very effective in restraining a sheep
and is one of the tricks of the shepherding
trade (Fig 22).

*Fig 22 Restraining a sheep. One hand
under a slightly raised chin will keep most
sheep in check.*

To handle a sheep:

1. Keep the group tightly mobbed.
2. Approach an individual from behind,
   where it has a blind spot.
3. Cup one hand under its chin and lift the
   head slightly above normal.

39

4. Grip the rump firmly with the other hand.
5. Walk it (preferably backwards) to the side of the pen and restrain with a hand under the chin and a knee against the flank.

Gently squeezing the tail encourages a sheep to move forward.

A sheep can be caught by grabbing the hind leg above the hock and lifting the leg off the ground, but this encourages a struggle.

Lifting

For trimming feet, shearing and sometimes dagging and crutching, sheep need to be upturned into a sitting position and it is essential to do this correctly to avoid back injuries. Light sheep are turned by lifting, and heavy sheep by unbalancing.

Sheep should never be picked up by the horns, fleece or legs and never grabbed by the wool. Try to avoid handling them when they have full rumens.

Fig 23 Turning a light sheep: Stand with both knees behind its shoulder and hold it under the jaw and by the flank (left) *then lift the hind leg and roll it over the knee* (left below) *and sit it upright on its rump* (below).

One way to turn a light sheep (Fig 23) is:

1. With one hand under the jaw, stand close with both knees behind its shoulder.
2. Grasp a fold of wool and skin low down on the far flank with the other hand.
3. Lift that hind leg off the ground, lean back and roll it over the knee onto its rump.
4. Turn it to sit between the handler's knees.

The shearers' catch is another method to turn a light sheep:

1. Approach from behind and grasp with the left hand around the neck and brisket.
2. Walk backwards lifting the front end quickly so that it rears up on its back legs.
3. Cross the right hand under the right foreleg and grasp the left foreleg above the knee.

4. Walk the sheep backwards to where it is wanted.
5. Take a longer backwards step, lift a bit more on the front and it will overbalance and sit on its rump.
6. Sit it up against the handler's knees but do not let its hind feet push against anything solid.

To unbalance a heavy sheep (Fig 24) the technique is:

1. Hold the sheep against the legs with one hand under the chin and the other on the rump.
2. Turn the head to look backwards along the shoulder and at the same time press down on the rump.
3. As it goes off-balance it sits down. Let it take a few steps backwards as it collapses.
4. Leave it on the ground or hold both upper forelegs and tilt it into a sitting position against the handler's legs.

Fig 24 Turning a heavy sheep: With one hand turn the head back to lie along its side and press on the rump (left) *with the other hand. It goes off-balance and collapses* (right). *It can be sat up on to its rump or left on the ground.*

Fig 25 Carrying a lamb.

The upturned sheep (Fig 23) is held at an angle of about 60 degrees – further forward and it gets back on its feet, further back and it struggles. Let the head fall to the right to give the rumen room to expand. It should not have anything to push against with its feet and because it is not comfortable sitting on its tail it should be eased onto a rump.

In some situations sheep can be left lying on the ground for working. A knee firmly on the neck and a sack over the head quietens them, but beware of flailing legs.

To carry a lamb (Fig 25), put one arm behind the forelegs and the other around the hindquarter.

Handling Aids

There are 'turnover crates' on the market (Fig 26) which are excellent but costly and if they are not part of a handling system it

Fig 27 The deck-chair – a useful piece of equipment which can be home built.

Fig 26 A turnover crate.

is difficult to get sheep into them. A simple and effective system is the deck-chair (Fig 27) which can be home built. The sheep is walked backwards towards it then tipped upright until it overbalances and sits down. It is then restrained by a strap so that the shepherd has both hands free to work.

Also available are yokes (Fig 28) which hold sheep by the neck for crutching and trimming, and immobilizers (Fig 29) for using in the field.

Handling in the Field

Catching an individual sheep in the field is useful for on-the-spot treatment. Tame sheep are no problem but wary sheep can be difficult. Sheep depend on flight for survival and at least one sheep in a resting flock will remain alert, so it is almost impossible to creep up on them. Some may be caught from the rear at the feed trough. Alternatively, a group may be cornered and temporarily contained by a length of wire netting secured at one end to the hedge or fence. If the netting is unrolled around them quietly and slowly the sheep are not frightened because they can see through it. Once the sheep are contained it is possible to catch a single sheep.

Fig 28 A yoke holds a sheep firmly by the neck and is an extra pair of hands for the single-handed shepherd.

Fig 29 A New Zealand invention, the immobilizer can be carried in the pocket and will restrain a sheep in the field while the shepherd goes for help, equipment or transport.

The Crook

The crook is the shepherd's badge of office and is the ultimate in stock control because it is an extension of the arm. The neck crook is better on shorn sheep but works on fleecy sheep and captures them safely and in a relatively immobile position. Take care when catching an animal alongside a fence that the crook does not catch the fence and the momentum of the sheep destroy the crook.

The leg crook is shorter than the neck crook with a smaller jaw but it must be used carefully to catch high on the hind leg to avoid breaking or twisting the leg and at the same time avoid trapping the udder.

Dogs

Sheepdogs are synonymous with shepherding. But having a sheepdog for the sake of it is a mistake because in most small flocks it is not necessary. The new shepherd who really wants a dog should buy a ready-trained five-year-old who knows the ropes; but remember that small flocks cannot provide enough work to keep a dog in good working condition.

Moving Different Classes of Sheep

When ewes and lambs are weaned and separated it is best to leave the lambs in the field and move the ewes. This is partly because it leaves the lambs in familiar surroundings, but also because a group of weaned lambs is not cohesive and is difficult to drive.

Heavily pregnant ewes need to be moved slowly but are quite manageable. The ones rushing ahead are probably barren.

Dry ewes are lively and quite often have a silly hour at dusk when they play.

Ewes with lambs at foot need very attentive moving (see earlier in chapter).

In the small flock there are not usually enough rams to form a flock, so they need to be handled as individuals.

HANDLING SYSTEMS

In its simplest form a handling system must allow a flock to be held and treated with minimum fuss. In the small flock the emphasis should be on easy handling rather than speedy handling. It should be designed to avoid the need to push, drag and carry sheep; they have legs and can walk.

The Site

Preferably choose a site which has:

- Free-draining soil.
- Easy access and exit.
- Shelter and some shade.
- Level or slight slope.
- Alongside an existing wall or building.
- Power and water available.

Design

The design should be simple and take into account:

- The total number of sheep, or the largest flock likely to be handled at one time.
- The jobs for which it will be used.

The gathering or holding pen is the reception area for the whole flock and is also used to hold sheep for long periods while waiting for the shearer or after veterinary treatment; on these occasions they should have access to water. It should be as large as practicable to make penning easy. Allow at least 0.5sq m per ewe, bearing in mind that a twenty-five ewe flock may include fifty lambs at certain times of the year.

The catching pen is the working area in which to dose, vaccinate, check udders, check teeth, condition score, treat an injury, tag, dag and trim feet. Catching pens should allow around 0.3sq m per sheep. Dimensions will vary with the size of the breed but capacity should be limited so that all the sheep are within easy reach and there is room to move and work.

A crush is a long and narrow (about 1.2m wide) version of the catching pen and holds sheep two abreast for easy working.

Avoid square pens – rectangular or funnel-shaped ones encourage the flow of sheep.

Pens are normally built from rails, netting or hurdles and the only solid divisions are where sheep move in opposite directions or in the 'race' (see later).

All surfaces and edges must be smooth, and rails and any cladding should be on the inside, with posts on the outside.

It should be possible to recycle sheep through the system for additional treatment or sorting, and there should be a separate exit. Gates must operate easily and securely.

The design should take account of sheep behaviour to encourage them to enter the pens and flow through them. Their preference is to:

- Go uphill.
- Go towards light.
- Go towards an opening or open space.
- Follow another sheep.

Temporary Handling System

Often the time and effort taken to move a few sheep through an elaborate penning system may not be worthwhile. Most jobs can be done by having a holding pen which can be reduced in size around the flock to become a catching pen (Fig 20).

By using hurdles a pen can be taken to the sheep and erected on temporary sites and because sheep get used to a penning and handling routine the layout should be similar at each site.

The pen should incorporate a gateway which offers a view of freedom and encourages the flock in.

Where a site is used regularly, a permanent funnel is useful and some strategically placed wooden posts to support the hurdles give extra security. Portable hurdles may lift

off the ground or be forced apart when under pressure from a flock. The hurdles should connect to each other at the top and bottom and should be easy to uncouple because, in this system, they operate as gates. Hurdles which use steel pins to connect them are the simplest to use and the best on uneven ground.

It is essential to have a stock marker to mark and identify each sheep as it is treated, and a pouch around the waist to carry the appropriate tools to the sheep instead of dragging each sheep to the tools.

There is a range of lightweight mobile handling systems on the market which incorporate a trailer, race and gates and allow customised layouts, but they are not cheap.

Basic Permanent System

A basic permanent handling system (Fig 21) comprises:

- A gathering or holding area.
- A catching pen or a crush.
- A race.
- A second holding pen.
- Perhaps a dip.

Construction

A free-draining hardcore base is all that is needed for most handling areas. Topsoil should be removed to a depth of about 200mm and refilled with stones to ground level. The first 125mm can be filled with clean hardcore or coarse gravel then covered with 75mm of 20mm rounded chippings or gravel screenings.

For concreted areas, such as the dip draining pen and some working areas, top soil should be removed, 150mm of hardcore laid and finished with a 100mm concrete mix of 1:2:4 (cement, sand, gravel).

Blocks can be used for solid walls but should be plastered and have no sharp corners. Timber posts and rails make secure fencing for the pens.

All gates should have sturdy latches and hinges to ensure their quick operation. Where gates are under pressure the latches should be at about half height.

The Components

The race holds sheep in single file, has solid sides and is 900mm high. It may incorporate a footbath, weigher or a drafting gate through which the sheep can be sorted into groups. Vaccinating, drenching and checking mouths are done by leaning over the top.

If the sides slope – around 0.6–0.8m at the top, narrowing to 0.3m at the ground – the race can accommodate sheep of mixed sizes while preventing them from turning around. However, a straight-sided race is easier to build and more versatile when incorporating equipment. The entrance should be disguised if possible by a slight curve or corner and the exit should be clearly seen as open space. To keep them flowing the race should hold six sheep at a time and be not less than 3m long.

A fibreglass removable footbath is easy to clean but if a concrete one is incorporated it can be covered with non-slip wooden boards when not in use. Footbaths are usually 150–200mm deep and where they touch the sides of the race there should not be a ledge for sheep to tiptoe along. A piece of old carpet on the bottom reduces splashing.

Modern foot treatments require sheep to stand in the liquid rather than walk through it, so sheep may stand in the footbath in batches rather than walk through in a continuous flow. Where there is no race, a stand-in footbath (Fig 30) can be accessed from the catching pen.

A second holding pen is handy for holding treated sheep for further treatment or for sorting into groups.

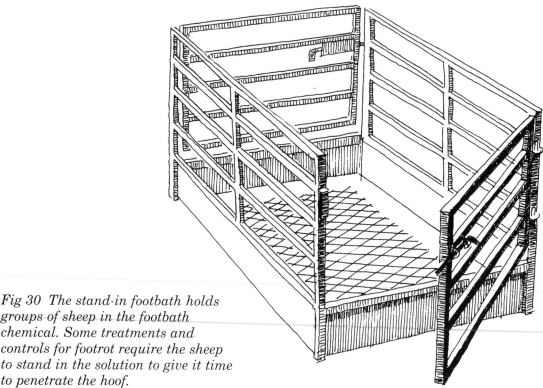

Fig 30 The stand-in footbath holds groups of sheep in the footbath chemical. Some treatments and controls for footrot require the sheep to stand in the solution to give it time to penetrate the hoof.

Sheep Dip

Dipping against skin parasites and fly strike is good husbandry, and dipping is important to pedigree flocks for show preparation. Controversy over the safety of the chemicals used to control external parasites has led to an increase in other control systems such as injections and pour-ons (see Chapter 8) but dipping is still widely practised in larger flocks.

Small fibreglass plunge dips take one sheep at a time and have a capacity of around 650 litres. There should be a concreted holding pen at one end and a draining pen at the other. A wet fleece can hold as much as 35 litres of liquid so the draining pen should hold sheep for five to ten minutes and have a 1:20 slope to allow liquid to drain back into the dip.

Each sheep is walked backwards to the edge and tipped in tail first. A board held across the bottom of the exit ramp stops them walking straight out. Other dipping options include hiring a mobile contract dipper or sharing a dip on another farm. Very small flocks manage by using a large oil drum, an old chest deep freeze or a domestic bath.

If dipping is not compulsory the small flock can be sprayed from a knapsack sprayer as it goes through the race.

There are regulations on the purchase, safe use, and disposal of dip chemicals so it is essential to check these with the Environment Agency and regional Ministry of Agriculture offices. Other regulations contain details of protective clothing and permits to buy dip chemicals.

TRANSPORT

Transport is essential on a sheep farm. It is usually cheaper to take a sheep to the veterinary surgeon rather than have him visit, and the sheep benefits from an equipped surgery. Sheep may also be transported to off-land grazing or in small numbers to market.

A small car or van with a tow hitch and a trailer are invaluable. A flat-bed trailer which converts to carrying stock (Fig 31) has many uses on a sheep farm. One measuring 1.5m × 2.5m will hold about 10–12 ewes. When moving ewes with baby lambs put the lambs in a partitioned area at the front of the vehicle; they encourage the ewes to get into the vehicle and are protected from being trampled during transit.

The general rules for transporting sheep are:

• Sheep travel best with an empty rumen.
• They are safer when reasonably closely packed.
• Larger vehicles may need a partition to reduce crowding or prevent falling.
• Have a non-slip, draining floor. Sand or an old carpet improve grip.
• Have plenty of ventilation but protect from the elements.
• There should be no exhaust fumes invading the area.
• They must not be able to fall or jump out.
• The transport must be covered.
• There must be no projections or sharp edges.
• There must be no gaps to trap limbs.
• There must be adequate headroom.
• Tethered and untethered animals should not be mixed.
• Sheep of different sizes should not be mixed.
• Keep watch in case any fall and risk suffocation.

Regulations

Flock owners transporting sheep locally will be subject to road transport regulations, welfare codes and common sense but may avoid the extra regulations which come into force when stock are transported for long distances. The police, farming organizations, the RSPCA, regional Ministry of

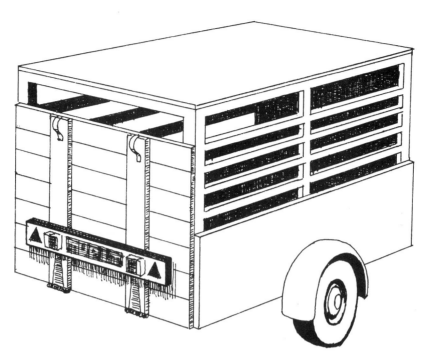

Fig 31 A stock trailer can be made by adding sides and a top to a low-sided trailer. A one-piece demountable top leaves the trailer available for general purpose use. The back hinges down to provide a ramp and the top could be plywood or lace-on canvas. It may be designed to utilize hurdles for the sides, but the gaps must be too narrow for the sheep to push their heads through and the base of the sides must be solid so that feet and legs cannot escape.

Agriculture offices, market operators and breed societies should be able to give appropriate advice.

Check the following:

- There may be a restriction on vehicle size and journey length (time and distance).
- Owners may need paper work for transporting, especially where there are restrictions due to disease control.
- A veterinary certificate may be needed to accompany sick or injured animals to a slaughterhouse.

Long distance transport (perhaps more than eight hours) is subject to extra regulations and it may be advisable to employ a professional haulier. Breeders who take their stock long distances to shows and sales will need to check the regulations.

Regulations may include:

- A portable ramp and side gates for loading and unloading.
- Restrictions on journey time or distance.
- Maximum length or size of pens in the vehicle.
- Rules on resting and feeding.
- Veterinary papers to accompany the load.
- Journey plan for the trip.
- Special training for drivers and assistants.
- Specific standards for the vehicles and trailers.

4 Breeding and Genetics ———————————

Breeding from the small flock needs planning and preparation.

The time to breed depends on:

- When the ewes are sexually active.
- When the lambs are wanted – to suit the market, the shepherd, the climate and the feed supplies (*see* Table 1 and Fig 103).

In temperate regions of the world sheep are seasonal breeders and will not mate until the daylight hours begin to shorten. With a gestation period of around five months (147 days) this ensures that there is spring growth of vegetation when the lambs are born.

The breeding season (when ewes are receptive to the ram) averages four months, starting in autumn and ending in winter. It is thought that it begins when the number of daylight hours fall below fourteen.

The seasonality is more pronounced in hill sheep than in lowland sheep and is stronger in the higher latitudes. Hill sheep may be sexually active for only two months while lowland sheep may breed for six to eight months. Some breeds, notably the Poll Dorset and Merino, are sexually active for at least eight months of the year and, when crossed with another breed, will pass on this trait. Rams are less seasonal than ewes in their breeding.

Ewes may be mated (tupped) any time when they are sexually active but their fertility is greatest during the middle of the breeding season. Fertility is also influenced by breed (the Finnish Landrace, Romanov and Cambridge are prolific lambers), by feed (good nutrition will stimulate ovulation) and by age (older ewes tend to be more fertile than young ewes). Peak fertility is reached at three to five years and maintained for another five years.

The Ram

The ram is a problem in small flocks. He is needed for only six weeks of the year but must have feed, care, and companionship while away from the ewes for the rest of the year. Where his daughters are kept for breeding, inbreeding becomes a possibility long before the end of his working life; in large flocks the average working life of a ram is three to four years.

Keeping two rams, or sharing two with another flock, solves the problem of companionship, insures against one becoming infertile and delays inbreeding. An alternative is to buy one or two ram lambs, breed from them and then sell them as two-tooths.

Borrowing is an option, but introducing disease is a risk for both the borrower and the lender. If borrowing, clarify who is responsible for any veterinary treatment, who is responsible if the ram dies and check that he is insured. See that he has no obvious signs of transmissible disease such as orf or foot rot.

Pedigree breeders often hire out rams but increasingly only to flocks with a guaranteed high health status. Hiring should guarantee a quality, healthy, fertile ram and avoid inbreeding. Clarify the period and the terms of hire such as veterinary costs, insurance and fertility guarantees. This latter is tricky because failure to breed may be the fault of the ewes and not the ram.

Sheep keepers who intend to keep their own replacement ewe lambs should consider using a ram which is scrapie genotyped to reduce the risk of introducing scrapie into the flock.

PREPARATION FOR BREEDING

In the new flock, preparation for breeding should begin at least eight weeks before tupping. In an established flock, preparation begins at weaning.

There are three stages:

- Selection and culling.
- Feeding.
- Veterinary treatments.

Selection

Selection is the choice of parents for the next generation. Stock for the new flock is selected at purchase, but in subsequent years some ewes will be culled and others brought in as replacements.

Selection should start well in advance of breeding so that culled stock can be sold and selected stock can be prepared.

Unless a flock is being expanded it is usual to cull from the oldest, bringing in youngsters to maintain mixed ages. A 10–15 per cent culling rate is normal in an established small flock.

Ewes to be retained in the flock are selected on:

- Physical soundness.
- How they perform.

Selection on physical soundness such as teeth, feet and udder are the same as described in Chapter 1 when buying ewes. However, the decision can now be modified by knowledge of the sheep. A ewe with a pendulous udder and an undershot jaw, but which reared fast-growing twins with no complications, could be worth keeping.

Provided the farming system is suited to elderly ewes (plenty of feed and personal attention) there is no need to cull on age. Ewes can breed to fifteen years.

A basic rule for selection is to keep ewes which produce a high total lamb weight at weaning. This selects for twins, milky mothers and good survival rate.

Reasons for which breeding ewes may be culled include:

- Barren for two years.
- Prolapsed at lambing.
- Damaged udder or suffered mastitis.
- Faulty teeth.
- Difficulty in rearing lambs.
- Problems conceiving and lambs late.
- Persistent dystokia (difficulties at lambing).
- Persistently lame.
- Bad mother.
- Poor milker.
- Poor fleece.

Reasons for culling a ram include:

- Suspect fertility.
- Health problems – especially bad feet.
- Teeth.
- Risk of inbreeding.
- Age.
- Aggressive to humans – to the point of danger.

Selecting Replacements

Culled ewes will be replaced either by bought-in ewes or by young ewes bred in the flock.

Keeping replacement ewe lambs from the flock has the advantage that their history is known, they are conditioned to the environment and they will not introduce disease. The disadvantage is that they need separate management from the adult flock until after their first lambing, which can complicate grazing management. It also complicates the use of rams to avoid ewes being bred to their own fathers.

Lambs kept for breeding should be identified early so that those not selected can be sold for slaughter.

Select on:

• Genetic quality.
• Physical quality.

Selection for genetic traits will depend on the purpose of the flock and will be based on the performance of the dam and any older siblings. Dairy flocks will look for high milk yields; wool flocks for fleece weight, colour and freedom from defects; meat flocks for lean, fast growing lambs with good conformation (*see* Chapter 11). Pedigree breeders will also be guided by the rules of their breed society.

Avoid lambs whose dams prolapsed, had bad feet or suffered persistent dystokia; and those whose dams or sires had scrapie or other heritable diseases.

Physical qualities include good feet, sound jaw and teeth, and a good fleece. Small teats and vulva in ewe lambs suggest breeding problems (hermaphrodites) and should be avoided. The circumference of the scrotum in ram lambs is related to fertility – the larger the circumference the more fertile he is likely to be (*see* Chapter 1).

Feeding

Ewes that are too fat or too thin (Fig 32) when they go to the ram may not breed. They may have fewer lambs (singles instead of twins) and have health problems later in pregnancy, such as pregnancy toxaemia. At the start of tupping, lowland ewes should be in condition score (CS) 3–3.5, upland ewes 2.5–3 and rams 3.5–4 (*see* Appendix II).

At eight to ten weeks before tupping, any thin ewes with CS2 or less should be put on good grass and fed up to 450g of compounds

Fig 32 Condition scoring a ewe. The way the fingers go under the transverse processes indicates that she is thin.

per head per day. Those at CS2–2.5 should improve on good grass alone. An improvement in condition from CS2.5 to CS3.5 represents a gain of about 6kg in body weight.

Fat ewes, that are CS3.5 or higher, should be kept on poor grass. Very fat ones (CS4–5) can be slimmed on a straw diet but protein pellets should be fed with the straw to prevent them losing protein as well as fat.

Splitting the flock into groups for separate feeding is not easy on a small farm but is worthwhile. It is important to handle the ewes regularly during this period to monitor their condition.

Flushing

Three weeks before tupping the ewes should be in ideal body condition and the whole flock put on good grazing for flushing.

Flushing (improving the nutrition of the flock) has the effect of increasing the ovulation rate in the ewes – possibly because it suggests to them that times are good and they can risk having extra lambs. But if extra lambs are not wanted – such as in hill ewes or ewe lambs, they should not be flushed but simply kept on continuous good feed. Some breeds are more responsive to flushing than others.

Flushing can be achieved on grass of around 6cm high, and supplementary feeding is only necessary for thin ewes or when grass height is less than 4cm. No ewes, not even fat ones, should be allowed to lose weight during this period. Rams also need flushing.

Veterinary Treatments

About three weeks before tupping the ewes should be crutched (Fig 33), their feet trimmed, dosed for internal parasites and vaccinated against clostridial diseases or other health problems (*see* Chapter 8).

Trace elements and vitamins such as vitamin E, selenium, manganese, zinc, iron and cobalt are important to reproduction. Where reproductive problems are experienced discuss feeding supplementary minerals with the veterinary surgeon.

Early lambing flocks, which are mated in the summer, must be shorn well before tupping or at least two months afterwards.

Preparing the Ram

Rams are described as being half the flock but in practice they are more. Along with their ability to improve the genetic quality of the flock, their virility and fertility influences the numbers of ova fertilized and, therefore, the number of lambs born.

Semen takes 6–8 weeks to produce so rams must be prepared for tupping at least eight weeks in advance. This should be borne in mind when they are hired or borrowed. Where appropriate they have the same veterinary treatments as the ewes, plus a physical check as described in Chapter 1.

Rams should be condition scored and fed similarly to the ewes. They must be in good condition at tupping (CS3.5–4) because they need stamina. In hot weather the wool can be carefully clipped from the belly and scrotum to keep the testicles cool.

Fig 33 Crutching a ewe before mating. Trimming the wool from around the tail is also important in the summer to prevent fly strike, before shearing to keep the wool clean and before lambing so that the lambing process is visible.

Between preparation and tupping the ram should not be stressed, and after mating he should not be banished and neglected until the next time he is needed.

TUPPING MANAGEMENT

A successful tupping should result in:

- All the ewes in lamb.
- A compact lambing period.
- Accurate lambing dates.

Tupping Timetable

At weaning	Select ewes for breeding.
8 weeks before tupping	Condition score ewes and rams and adjust feeding. Check ram fertility, prepare feet, give veterinary treatments.
3 weeks before tupping	Veterinary treatments for ewes. Flush adult ewes and rams.
About 2 weeks before tupping	Synchronize with teaser ram or hormonal sponges (if appropriate).
7 days before tupping	Fit ram harness (if appropriate).
2 days before tupping	Remove sponges.
Tupping	Remove teaser ram, put in raddled ram. Flock on good grass and left undisturbed.
15 days after tupping	Change raddle colour (earlier if 50% of the flock has mated).
30 days after tupping	Change raddle colour.
5 weeks after tupping	Change colour of raddle or take rams out.
12–14 weeks after tupping	Pregnancy test by ultrasonic scanning.

Fig 34 A ram shows his interest in a ewe which is on heat. Pawing her flank and curling his lips (flehmen reaction) are typical behaviour.

Tupping should be on good pasture because it is vital that sheep do not lose weight during and immediately after tupping. Grass height should be kept at around 6cm (*see* Chapter 6). A useful figure to assist planning is that an 8cm high pasture will carry twelve ewes per hectare for two breeding cycles (about a month).

Rams may need 450g of an 18 per cent protein concentrate each day to maintain stamina and libido and the whole flock may need feeding if there is persistent heavy rain or grass supplies become short. Hand feeding the ram gives a chance to check him, but some shepherds are opposed to this because it distracts him.

In hot weather there should be shade because heat depresses sexual activity.

In a flock which has not been synchronized to bring the ewes on heat at the same time, one mature ram can serve forty ewes.

Ewes come on heat (oestrus) every 15–17 days, it lasts for twenty-four to thirty-six hours and ovulation occurs at the end of the heat. Sheep are not demonstrative when they are on heat and the only indication will be the attention of the ram (Fig 34). Approximately 6 per cent of the flock is served each day during the first seventeen days and about 80 per cent of the ewes should be pregnant at the end of this time.

Ewes that are pregnant after the first ovulation are described as having 'held to first service'. Those that do not hold to first service but get pregnant at their next ovulation will have 'held to second service'.

Rams stay in the flock for around six weeks so that every ewe can have two or three ovulations and therefore two or three chances of becoming pregnant. If a second group of ewes, such as ewe lambs, are tupped after the main flock then any adult ewes that have failed to get pregnant can be put in with them for another chance. A record of tupping (*see* box) is vital to identify fertility problems and for flock management.

Tupping Records

Number of ewes in group:
Identity of ewes:
Identity of rams:
Ram to ewe ratio:
Weather conditions:
Date ram in:
Date ram out:
  Date and colour of raddle change:
    1st
    2nd
    3rd
    4th
Dates when rams changed/substituted:
The identity (and dates) of the ewes marked
  for 1st service:
The identity (and dates) of the ewes marked
  for 2nd service:
The identity (and dates) of the ewes marked
  for 3rd service:
The identity of any unmarked ewes:
Number and percentage holding to 1st service:
Number and percentage holding to 2nd
  service:
Number and percentage holding to 3rd service
Date for pregnancy scanning:
Predicted start of lambing:

This is the basic information that a small flock keeper should record during and after tupping

---

RADDLING

To indicate what is happening at tupping, a ram wears a coloured marker (raddle) on the brisket so that when he mounts a ewe she is marked on the rump and identified as being mated. The colour of the raddle is changed at least every fifteen days, thus when a marked ewe is marked for a second time it is clear that she did not hold to the ram at the previous mating. It is acceptable for around 15 per cent to fail to hold to their first service and a few may go to three matings. Where a high proportion of ewes are re-marked every seventeen days it will cast suspicion on the fertility of the ram and he should be replaced immediately.

The final mark will indicate when the ewe was successfully mated and, therefore, when she will lamb. The lambing date will be about 147 days after the service (*see* Table 2). A ewe that is unmarked at the end of tupping is not necessarily barren; in wet or cold weather the crayons may fail to mark. Or she may already be pregnant from a precocious ram lamb before weaning or from a ram which has broken into the

Table 2 Gestation Period for Ewes

Tup	Lamb	Tup	Lamb	Tup	Lamb	Tup	Lamb	Tup	Lamb	Tup	Lamb
Jan	May	Feb	June	Mar	July	Apr	Aug	Tup	May	Sept	Jun	Oct
1	28	1	28	1	26	1	26	1	25	1	26
2	29	2	29	2	27	2	27	2	26	2	27
3	30	3	30	3	28	3	28	3	27	3	28
4	31	4	Jul 1	4	29	4	29	4	28	4	29
5	Jun 1	5	2	5	30	5	30	5	29	5	30
6	2	6	3	6	31	6	31	6	30	6	31
7	3	7	4	7	Aug 1	7	Sep 1	7	Oct 1	7	Nov 1
8	4	8	5	8	2	8	2	8	2	8	2
9	5	9	6	9	3	9	3	9	3	9	3
10	6	10	7	10	4	10	4	10	4	10	4
11	7	11	8	11	5	11	5	11	5	11	5
12	8	12	9	12	6	12	6	12	6	12	6
13	9	13	10	13	7	13	7	13	7	13	7
14	10	14	11	14	8	14	8	14	8	14	8
15	11	15	12	15	9	15	9	15	9	15	9
16	12	16	13	16	10	16	10	16	10	16	10
17	13	17	14	17	11	17	11	17	11	17	11
18	14	18	15	18	12	18	12	18	12	18	12
19	15	19	16	19	13	19	13	19	13	19	13
20	16	20	17	20	14	20	14	20	14	20	14
21	17	21	18	21	15	21	15	21	15	21	15
22	18	22	19	22	16	22	16	22	16	22	16
23	19	23	20	23	17	23	17	23	17	23	17
24	20	24	21	24	18	24	18	24	18	24	18
25	21	25	22	25	19	25	19	25	19	25	19
26	22	26	23	26	20	26	20	26	20	26	20
27	23	27	24	27	21	27	21	27	21	27	21
28	24	28	25	28	22	28	22	28	22	28	22
29	25			29	23	29	23	29	23	29	23
30	26			30	24	30	24	30	24	30	24
31	27			31	25			31	25		

Table 2 Gestation *(continued)*

Tup Jul	Lamb Nov	Tup Aug	Lamb Dec	Tup Sep	Lamb Jan	Tup Oct	Lamb Feb	Tup Nov	Lamb Mar	Tup Dec	Lamb Apr
1	25	1	26	1	26	1	25	1	28	1	27
2	26	2	27	2	27	2	26	2	29	2	28
3	27	3	28	3	28	3	27	3	30	3	29
4	28	4	29	4	29	4	28	4	31	4	30
5	29	5	30	5	30	5	Mar 1	5	Apr 1	5	May 1
6	30	6	31	6	31	6	2	6	2	6	2
7	Dec 1	7	Jan 1	7	Feb 1	7	3	7	3	7	3
8	2	8	2	8	2	8	4	8	4	8	4
9	3	9	3	9	3	9	5	9	5	9	5
10	4	10	4	10	4	10	6	10	6	10	6
11	5	11	5	11	5	11	7	11	7	11	7
12	6	12	6	12	6	12	8	12	8	12	8
13	7	13	7	13	7	13	9	13	9	13	9
14	8	14	8	14	8	14	10	14	10	14	10
15	9	15	9	15	9	15	11	15	11	15	11
16	10	16	10	16	10	16	12	16	12	16	12
17	11	17	11	17	11	17	13	17	13	17	13
18	12	18	12	18	12	18	14	18	14	18	14
19	13	19	13	19	13	19	15	19	15	19	15
20	14	20	14	20	14	20	16	20	16	20	16
21	15	21	15	21	15	21	17	21	17	21	17
22	16	22	16	22	16	22	18	22	18	22	18
23	17	23	17	23	17	23	19	23	19	23	18
24	18	24	18	24	18	24	20	24	20	24	20
25	19	25	19	25	19	25	21	25	21	25	21
26	20	26	20	26	20	26	22	26	22	26	22
27	21	27	21	27	21	27	23	27	23	27	23
28	22	28	22	28	22	28	24	28	24	28	24
29	23	29	23	29	23	29	25	29	25	29	25
30	24	30	24	30	24	30	26	30	26	30	26
31	25	31	25			31	27			31	27

Gestation period for ewes averages 147 days (five months) and can vary from 142 to 150. Breed of ewe rather than sexes or size of litter seem to govern the period. Spring-born lambs may be carried longer than autumn born, and pregnant ewes that are winter-shorn may carry lambs for one day longer than those that are not shorn. Make an allowance for Leap Years.

flock. The latter can result in 10 per cent conception overnight.

In the small flock it is possible to keep a daily record of the ewes being tupped so that their expected lambing date is accurate – a bonus at lambing time.

Markers may be applied as a paste, painted on the brisket each day. The ram must be easy to catch, and if he is shorn some wool should be left on the brisket to hold the paste. Alternatively, a harness (Fig 35) is buckled around the chest to hold a coloured crayon. Harnesses should be fitted up to a week before tupping and should be watched carefully for chafing, especially under the forelegs. The straps may need tightening as they bed into the wool and as the ram loses weight during tupping.

Raddle colours are used in sequence starting with the lightest (white or yellow) and progressing through the darker colours from orange, red, blue, green and black. When running two or more rams together it helps to identify any that are not very active by raddling them with different colours, but it will not identify the sires of the resulting offspring because most ewes will have been served by all the rams. It is a characteristic of the ewe that she can be fertilized by more than one ram and give birth to twins with different fathers.

Ram Behaviour

There is a saying that rams should be run in threes – two to fight and one to get on with the job. Rams will fight over ovulating ewes and reduce the chances of successful matings. A dominant ram will get all the business, but it will be a disaster if he is infertile.

Where two rams are used it is best to split the flock and run one ram with each group. They can be swapped every few days in case one is infertile or is concentrating his attention on a favourite ewe. The flocks should be out of sight of each other otherwise each ram will be distracted by the other.

It is thought that unshorn ewes are more attractive to rams than shorn ones because wool carries pheromones (odours to attract a mate). Sheep which have been recently dipped may not be readily mated. When given a choice, rams show a preference for ewes of their own breed. They may also create harems.

Fig 35 A crayon in a ram harness will mark ewes as they are mated and is a reliable indicator of what is happening at tupping. Harnesses should be fairly tight but not chafing and checked regularly. Attaching a harness to an unwilling ram is less of a puzzle if the numerous buckles and straps are pre-marked to indicate which goes with which.

Rams can be aggressive, especially during the mating season. Beware of turning your back on them, especially in a field, and be cautious about putting them in fields where there are public rights of way. A slap on the nose checks them, but persistently aggressive rams are dangerous and should be culled.

Strange rams may fight each other to the death, even through fences – leading to court cases between neighbours.

When strange rams are run together they should first be confined for a day in a space too small for them to run at each other. An American trick is to hang old tyres at ram shoulder height to prevent them getting a clear run.

Ram Lambs

Rams reach sexual maturity by six to nine months and can be used as yearlings. They should be run with small groups of mature ewes – about one ram lamb to twenty to twenty-five ewes – and not used with young ewes. They should not have to compete with older, dominant rams.

Ewe Lambs

Ewe lambs need good management if they are bred in their first year because they are growing their own bodies as well as those of their lambs.

There are, however, the following advantages:

- They subsequently make better mothers.
- They do not get too fat.
- The first lamb is a financial bonus.

Ewe lambs do not usually start their breeding cycle until they are 8–10 months old, but puberty in many breeds is also influenced by weight. There is a saying that 'if they are big enough they are old enough', but they should be around 65 per cent of their mature weight before breeding. Mature weight is half the combined adult weight of dam and sire. Around 30 per cent may not breed in the first year.

Ewe lambs are unresponsive to the ram and need a persistent, active adult ram. The ram to ewe ratio is usually 1:30 and ewe lambs may be tupped a month after the adults to give them extra time to grow, have extra attention at lambing and to extend the use of the ram. Rams should be run for only two oestrus cycles (about five weeks) to reduce the lambing period. If adult and young ewes are tupped at the same time they should be kept in separate groups.

Shearing ewe lambs in late summer can improve their growth rate. To discourage large litters they should not be flushed but kept growing steadily on good feed.

Compact Mating

A short mating period means a short lambing period. Short lambing periods may not be welcome in a large flock where resources will be stretched, but in a small flock the advantages are:

- Easier to feed the pregnant ewes accurately.
- Less wasted time and sleepless nights waiting for ewes to lamb.
- More chance of fostering lambs on to newly lambed ewes.
- Lambs are easier to market because they are similar ages.

A compact mating is the result of the flock being synchronized so that the ewes come on heat at the same time. Two common aids to synchronize a flock are:

- A teaser ram.
- Hormone-impregnated sponges.

A teaser ram is an active ram which has been vasectomized by the veterinary surgeon to make him infertile but still produces the male pheromone to stimulate the female – what is known as the 'ram effect'. His role is to synchronize oestrus in the flock by stimulating ewes to come into heat. He can also bring forward the breeding season by about three weeks.

Teaser rams must be strong, active rams with good libido. A vasectomized old ram which is past his sell by date is unlikely to be effective. A young entire ram lamb that missed the market or may have been a hand-reared favourite could be a candidate. Some pedigree breeders vasectomize and sell rams that they do not want to breed.

The ewes should be out of sight, sound and smell of rams for at least six weeks prior to the teaser going in. The teaser is introduced suddenly about seventeen days before the planned mating. One teaser per fifty ewes is adequate. He is removed after sixteen days and replaced by the fertile ram. Most ewes have their first true heat and are mated during the first week.

Sponges

Teaser rams give only partial synchronization. A more accurate system is the use of progesterone sponges that are inserted in the vagina of the ewe sixteen days before mating (Fig 36). Progesterone is the hormone which is released naturally during pregnancy to inhibit the breeding cycle.

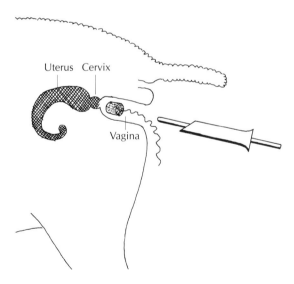

Fig 36 Sponges are inserted in the vagina to synchronize mating. The sponge is loaded into an applicator which is inserted into the vagina and the sponge deposited against the cervix. The sponge is held in position with a rod while the applicator is withdrawn and the strings are left exposed so that the sponge can be pulled out two weeks later.

The progesterone (or the analogue progestagen) which is released by the sponge mimics pregnancy and arrests ovulation. When the sponges are withdrawn after fourteen days the body reacts to the drop in progesterone and starts to cycle; the result is that all the treated ewes come into heat at the same time. Rams are put in forty-eight hours after sponge removal and tupping lasts for about three to four days. To make his life easier the ram is not raddled for the first heat period but raddled for the second; any unmarked ewes are assumed to have held to first service.

With this technique there is extra demand for ram power and they are used at a ratio of one ram per eight ewes – which, in the small flock, requires borrowing. Ram power can be reduced to one per twelve ewes by operating a 'hand mating' system. In this system the ram and his ewes are penned together forty-eight hours after sponge removal (timed to be in the morning) and as each ewe is seen to be correctly served she is removed from the pen. At the end of the day, or when all the ewes have been served, they are all run together. The system prevents rams from concentrating on a favourite ewe and improves the percentage of ewes holding to first service.

Ewes holding to first service (around 60–80 per cent) will lamb down over a six-day period followed by a break of about ten days before the ewes holding to second service start lambing. This break gives the shepherd a chance to catch up on sleep and to attend to the first group of lambs.

AFTER TUPPING

Up to 30 per cent of eggs shed at ovulation may fail to develop into lambs.

This loss is due to:

- Eggs failing to be fertilized.
- Fertilized eggs failing to implant.
- Foetuses being reabsorbed.

Stress – due to rough handling, poor nutrition, changes in nutrition, dogs or extremes of weather – is the main cause of these losses.

After fertilization the egg (ovum) floats in uterine fluid until it attaches (implants) to the uterine wall three weeks after conception. During this period nutritional or physical stress can prevent implantation and the ovum is lost.

After a successful implantation, stress in the early stages of pregnancy can cause the ewe to reabsorb a foetus. If this occurs early in the mating period a ewe can still re-mate. To minimize these losses the flock should remain on good grass and not be disturbed from the time that the ram is put in to at least one month after he is taken out.

Newly served ewes must have shade and water because the new conceptus is sensitive to heat stress for its first two weeks.

Pregnancy Testing

Ewes can be tested to see if they are pregnant. On-farm testers are available that simply indicate whether or not a ewe is in lamb.

More accurate results come from ultrasound scanning that not only indicates pregnancy but also the number of lambs each ewe is carrying. This allows accurate feeding and management during pregnancy, identifies barreners early and is useful information at lambing.

The job is done by contractors, ideally 12–14 weeks after the beginning of tupping. Although it is a useful tool, scanning may not be economic for the small flock.

AIDS TO BREEDING

Ewes can be persuaded to breed outside their natural breeding period, usually in order to produce lambs for the early market, to advance lambing so that it does not clash with other farm activities, or to have pedigree lambs well-grown by sale time.

A natural way to advance the breeding season is to introduce a breed with a long season such as the Dorset Horn and Poll Dorset; just a quarter of their blood in another breed can extend its breeding period.

Another technique involves the use of melatonin, a naturally occurring hormone produced by the pineal gland as a response to alternating night and day. It is also used to combat jet lag in humans.

In sheep it is administered as a subcutaneous implant that creates a chemical darkness, convincing the ewe that the daylight hours are shortening. It can bring the breeding season forward by up to six weeks. Teaser rams are put in thirty-five days after implantation and the entire rams seventeen days after that. Peak matings occur around fifty-eight days after implantation.

The breeding season can be advanced by up to six weeks by an injection of Pregnant Mare Serum Gonadotrophin (PMSG) after the withdrawal of progesterone sponges used for synchronization. This also stimulates ovulation, sometimes causing big litters.

Artificial Insemination

The small flock owner, especially the pedigree breeder, could consider artificial insemination (AI). It is not an easy alternative to natural mating but it allows breeders to use top rams without the expense of owning them. However, it does not eliminate the need to keep a ram because one is necessary to run with the flock after insemination to serve ewes that did not conceive to AI.

AI involves taking semen from a ram and depositing it into either the cervix or the uterus of the ewe.

Semen is collected from a ram in an artificial vagina while he is mounting an on-heat ewe. Best results come from using it fresh but frozen semen is more convenient. Owners of top rams often have semen frozen and stored as a safeguard against death.

Fresh semen is usually diluted and kept at a constant temperature during insemination.

The cervix of the ewe is tortuous (*see* Appendix I, Fig 114) and the semen cannot be deposited right into the uterus – only into the first fold of the cervix. Best results – 70–75 per cent conception – come from laparoscopic AI (LAI) in which the semen is deposited surgically into the uterus. Unlike cervical AI this must usually be done by a veterinary surgeon.

Ewes must be inseminated at the right stage of oestrus and commonly they are synchronized using progesterone sponges.

Flock owners can learn cervical AI and set up as operators. AI companies which advertise in the agricultural press will give advice.

Table 3 Average Percentage of Heritability	
Birth weight	30
Eight week weight	18
Weaning weight	30
Mature body weight	40
Growth rate	25
Fleece weight	30
Fleece quality	45
Multiple births	15
Milk production	20
Carcass conformation	35

Embryo Transfer

The elite of all the breeding techniques is embryo transfer (ET) where eggs from superior ewes are fertilized by sperm from top rams and implanted in recipient ewes who give birth to the offspring. The technique accelerates breed or flock improvement and is only appropriate to serious pedigree breeders.

BREEDING FOR IMPROVEMENT

The quality of a flock is maintained or raised by selecting and breeding from the best stock.

Success is influenced by:

- The number of sheep.
- The generation interval.
- The accuracy of flock records (to identify the best sheep).
- The heritability of the traits (Table 3) being selected for.

Breeders need a clear vision of what they want to achieve and should select for a minimum number of traits at one time.

Some traits are more heritable than others and it is difficult to identify whether differences are due to genetics or to environmental influences such as nutrition or climate. For example, Table 3 shows that in a group of lambs of the same breed the differences in birthweight are 30 per cent inherited. The other 70 per cent is due to environmental factors such as ewe nutrition and size.

The table also shows that multiple births, with an average heritability of 15 per cent, are not highly influenced by genetics. It is quicker to use a ram from a prolific breed such as the Cambridge or Finnish Landrace and breed prolific progeny in one generation.

Growth rate is a more heritable trait. It can be influenced by selection and justifies the use of a high growth rate ram. Wool quality is also readily heritable but is also influenced by nutrition and climate.

Introducing a new breed or just a new breed of ram is a quick route to genetic improvement but the breed must be suited to the environment and farming system. The improvement must also be beneficial; breeding for prolificacy on a harsh hill farm would be a disaster.

Crossbreeding

A quick way to improve a commercial flock is to introduce crossbreeding. This is the mating of animals of different breeds.

61

Crossbreeding is used to:

* Introduce new or improved characteristics into a flock – such as conformation, improved wool or milkiness.
* Create hybrid vigour or heterosis.

The performance of the crossbred animal is usually superior to the performance of each purebred parent. Crossbred ewes such as the Mule, Masham, Scotch Halfbred and Welsh Halfbred represent more than half of the UK breeding flock.

The hybrid vigour generates fast growth, disease resistance and larger lambs which reach sexual maturity sooner. It usually promotes the traits which are least responsive to selection, such as lamb survival. Mating crossbred ewes to a third breed to produce slaughter lambs gives maximum benefit from hybrid vigour, and ewe lambs from this third cross make good breeding stock.

When a cross is stabilized – crossbred animals bred with crossbred animals – the hybrid vigour is lost, and some sheep revert in type to the original purebred stock.

Crossbred rams also benefit from hybrid vigour. It has been shown that the combined effect of breeding crossbred rams with crossbred ewes was a 20 per cent increase in the number and weight of weaned lambs.

Inbreeding

Inbreeding is the mating of closely related animals and is used to improve or to eliminate certain characteristics. Closebreeding and linebreeding are types of inbreeding.

It can produce the uniformity sought by pedigree breeders but a common effect of inbreeding is to reduce fertility and growth rate and to increase lamb mortality. It may occur in numerically small breeds where unrelated stock are difficult to find and in breeds which become popular and are bred quickly to meet demand. A flock which sees

an unexplained increase in defects plus a decline in production may be suffering from inbreeding.

Intentional inbreeding should only be done in flocks which are already superior and are large enough to stand the necessary hard culling. Average or below average purebred flocks make better progress by introducing an unrelated superior ram.

Breeding Schemes

Efforts to improve sheep genetically have spawned schemes in which breeders can make objective measurements of their breeding stock and select the best. In essence there are two types of improvement programme in the UK – Group Breeding Schemes and Sire Reference Schemes.

The principle of Group Breeding Schemes is that breeders contribute high-quality ewes to a nucleus flock and select outstanding rams to sire successive generations. The measurements that are used to provide an overall selection index (an indicator of genetic merit) vary from breed to breed but generally they include mature size, lamb growth rate, eye muscle depth, back fat depth, litter size, and wool weight and quality.

Sire Reference Schemes were originally developed to improve the growth rates and muscularity of terminal sires but have been adapted to improve other important characteristics. The schemes create genetic links between flocks; the links are provided by having groups of lambs of similar breeding in each participating flock. These link lambs are the progeny of small teams of reference rams which are mated to a proportion of ewes in each flock; they are half-sibs and are used as a benchmark against which the genetic merit of all other lambs in the participating flocks are measured. Once the performance of every lamb has been measured and compared, outstanding lambs are selected for breeding.

5 Nutrition ————————

Nutrition is the single most important aspect of flock management. Poor nutrition underlies most of the common disease and performance problems affecting sheep.

Sheep derive 90 per cent of nutrients from forage, but most flocks need some supplementary energy, protein and minerals at specific times.

Sheep are ruminants and their digestive system comprises four stomachs (*see* Appendix I, Fig 113) which enable them to perform two important functions:

- To swallow food quickly then withdraw to safety to regurgitate and chew it (cud) at leisure.
- To digest and derive energy from the fibrous cells of plants

The swallowed grass goes into the rumen/reticulum and, at a convenient time, is regurgitated as a cud and chewed into smaller pieces. This process stimulates and incorporates saliva in the food (a sheep can secrete around 10 litres of saliva a day) to aid digestion. It is estimated that a sheep spends a third of its time chewing its cud.

Cudding usually starts about half an hour after the animal has finished eating and sheep should be allowed two hours' rest after feeding to allow cudding to complete, otherwise disturbance and exertion can cause stomach disorders.

The cuds go back into the rumen and are broken down and fermented by microbes to produce a mixture of fatty acids; these are an important source of energy. A by-product of this process is gas which must be expelled from the rumen by belching, otherwise it can result in the potentially fatal condition of bloat.

The liquified rumen contents then go to the reticulum and on to the omasum which is capable of grinding any remaining fibres and absorbing water. The processed feed passes to the abomasum – which is often called the true stomach – where enzymes are produced to digest sugars and starch, and acidity kills any surviving micro-organisms. The material then moves to the intestines where remaining nutrients and water are absorbed.

The practical implications of this system are:

- Feeding the sheep is an exercise in feeding microbes.
- The microbes need time to adapt to new feed – so this must be introduced gradually.
- Sheep must always have fibre. It is needed for the mixing action of the rumen and to stimulate the production of saliva.

Energy

The major nutritional requirements are energy, protein, minerals, vitamins and water.

Low energy is the most common nutritional problem. Energy is used for maintenance – to keep the body working – and for production such as growth, pregnancy and lactation, and is stored as fat. An insufficient energy supply will result in the ewes using up their fat reserves and losing condition. Roughage, like grass, hay, silage and root crops, meet most energy needs; concentrate feeds (see later) are generally high in energy and are fed to supplement low-energy rations.

Energy is expressed in megajoules (MJ), but not all the energy in a feed is used by the sheep. The amount which is used is expressed as the Metabolizable Energy (ME) and this is defined as the number of megajoules per kilogram of dry matter (MJ/kgDM). The dry matter is the material left after all moisture has been removed by heating.

The ME of good quality pasture, for example, (Table 4) is 12.5MJ/kgDM; this means that every kilogram of dry grass contains 12.5 megajoules of available energy. The higher the ME figure the more dense or concentrated is the energy in the feed.

High energy feeds are usually necessary for:

- Ewes in late pregnancy and lactation.
- Lambs needing to grow quickly to catch the market.
- Improving the condition of ewes and rams before mating.

The energy needs of ewes double in late pregnancy and treble in early lactation. As a guide the requirements of a 60kg (small) lowland ewe are:

Maintenance	8	MJ per day
Late pregnancy*	16	MJ per day
Early lactation*	26.5	MJ per day

Expecting twins and losing 0.5 CS.

A dry ewe gets enough energy from grass to maintain her, but the same sheep carrying twin lambs will not (Table 5). By knowing

Table 4 Composition of Common Sheep Feeds

	Dry Matter %	ME MJ/kgDM	CP g/kg/DM	% CP in DM	Av. degradability of protein %
Barley	86	13	120	12.0	90
Oats	86	12.1	105	10.5	90
Sugar Beet					
Pulp (dried)	90	12.5	128	12.8	75
Soya Bean Meal					
(extracted)	90	13.3	520	52.0	70
Fishmeal	90	14.2	700	70.0	35
Fodder Beet	13	12.5	60	6.0	85
Swedes	12	14.0	91	9.1	90
Molasses	75	12.8	44	4.4	80
Good hay	85	10.0	145	14.5	75
Poor hay	85	7.5	90	9.0	75
Barley straw	86	7.0	38	3.8	80
Good silage	25	11.5	145	14.5	80
Good pasture	25	12.5	156	15.6	75
Poor pasture	40	8.0	80	8.0	75

This table gives an indication of the energy and protein values of various feeds as well as the degradability of the protein.

both the ME of rations and the daily ME requirements of a ewe, she can be fed the appropriate amount of feed.

For example, a shepherd has bought a load of fodder beet. From Table 4 the ME of fodder beet is 12.5 and the dry matter 13 per cent. How much should he feed to a ewe which needs 10 MJ ME per day? The formula is:

$$\frac{MJ/day \times 100}{ME/kgDM \times DM\%} = \frac{10 \times 100}{12.5 \times 13} = \frac{1000}{162.5} = 6.2$$

He needs to feed 6.2kg of fresh fodder beet. A ewe's average daily appetite is around 1.0–2.0kgDM and, at first sight, 6.2kg of fodder beet may seem a lot to feed; but at 13 per cent DM it is only 0.8kgDM. On a practical note, fodder beet should not comprise more than 60 per cent of a ewe's daily dry matter intake (*see* Chapter 6).

This is a useful guide when planning how much foodstuff to grow, buy or feed but it has to take into account that a ewe has a limited appetite and may not be able to physically consume enough low dry matter (watery) feeds such as roots, or low ME value feeds such as poor hay, to get enough energy.

Protein

Protein builds tissue and is vital to the pregnant and lactating ewe and the growing

Month	Activity	Energy needs	Feed		Ideal CS
		MJ of ME	*Roughage*	*Concentrates*	
1	Flushing	12.5	Good grass 7cm	For thin ewes only	3.0
2	Tupping	12.5	Good grass 7cm	For rams	3.5
3	1st month pregnant	12.0	Good grass 7cm		3.5
4	2nd month pregnant	8.5	Grass 4–5cm		3.5
5	3rd month pregnant	8.5	Grass + 1.5kg hay	150g/head/day starting 8 weeks before lambing	3.5
6	4th month pregnant	15.0	1.5kg good hay or 4–5kg silage	and increasing by100g weekly to peak at 900g	3.5
7	Lambing	19.0	1.5kg good hay or 4–5kg silage	at lambing. Decrease until after lambing 6	3.5
8	Lactation	24.0	Grass 4cm plus 1.5kg good hay	weeks or when adequate (7cm) grass	3.0
9	Lactation	16.0	Good grass 7cm		2.5
10	Lactation Weaning	16.0	Good grass 7cm		2.5
11	Dry	8.0	Stale grass		2.5
12	Dry	8.0	Grass 5cm		2.5

Table 5 Basic Feeding Guide

This is for 70kg spring-lambing lowland ewes likely to be carrying twins and will be modified for: small ewes, ewes with singles, where good quality silage is fed, and whether the flock is outside or housed. A housed ewe may need 1.2MJ per day less than an outside ewe. In late pregnancy and early lactation, concentrates will need to be around ME 12.5 and 16–18 per cent CP. Roots can replace up to one third of concentrates on a dry matter basis up until two weeks before lambing.

lamb. Unlike energy, sheep do not store protein to draw on in times of shortage.

The protein content of feeds is described as the percentage of crude protein (CP) in the dry matter. For maintenance a ewe needs a feed which contains about 12 per cent CP in the dry matter, and growing and pregnant sheep need 16 to 18 per cent. The latter is higher than is found in average grassland but when they graze, sheep select out vegetation, such as clover, which has a high protein content.

Again, when feeding protein it is the rumen organisms which are being fed. There are two types of protein – Rumen Degradable Protein (RDP) and Undegradable Protein (UDP). The former, such as proteins in grass, are broken down quickly to ammonia in the rumen and used by the microbes. When the microbes die this microbial protein is absorbed by the animal. Raw materials which are rich in UDP such as fishmeal and heat-treated soya bean meal, escape being broken down in the rumen and go direct to the intestines where it is absorbed as amino acids.

In simple terms RDP feeds the rumen organisms and, indirectly, the sheep; UDP supplies protein to the sheep direct. Microbial protein can maintain the animal and UDP is only needed for growth and production in late pregnancy and early lactation. It also improves wool growth.

Minerals and Vitamins

Mineral deficiencies in grass are usually the result of deficiencies in the soil. If there is a local problem the veterinary surgeon will know and advise on correction. Symptoms of mineral deficiencies are often chronic rather than acute, such as poor growth and wool, poor fertility, loss of appetite and diarrhoea – all of which can have other causes. Soil and herbage tests will reveal deficiencies as will blood tests on the animals.

Major minerals needed are calcium, phosphorus and magnesium. Those needed in smaller quantities, known as trace elements, include iodine, cobalt, selenium, iron, manganese, molybdenum and zinc. The availability of some of these elements can be affected by the acidity of the soil (see Chapter 6).

Minerals are usually fed direct to the animal via mineral licks or in the feed. Injections or boluses may be advised by veterinary surgeons in certain cases. Deficiencies can be complex because an excess of one mineral may prevent the uptake or the utilization of another. For this reason blanket feeding a cocktail of minerals may not solve a particular problem but should reduce the risk of deficiencies.

Generally sheep are intolerant of copper, and since pig and poultry manures are high in copper (originally added to the feed) they should not be put on sheep grassland. Cattle and calf feeds normally contain added copper and must not be fed to sheep.

Calcium and magnesium are especially important at around lambing and lactation (see Chapter 9) and cobalt and selenium are important to the growing lamb.

Vitamin deficiency is uncommon in grazed sheep but vitamin A which is derived from green feed can be low when sheep are not on grass or are on old hay. Both vitamins A and D may need to be fed to housed sheep. Vitamin E is thought to be important to pregnant ewes to reduce lamb mortality. Most concentrate feeds which are formulated for sheep are balanced for vitamins and minerals.

Water

Sheep must always have access to clean water although they do not always drink it. Fresh grass can provide enough for dry sheep which means that water troughs may become stagnant and need regular cleaning. Allow five to eight litres of water per sheep per day in hot weather (sheep control their temperature by sweating and panting) and during pregnancy and lactation. Very cold water can have an adverse effect on the rumen bacteria and affect digestion.

Concentrate Feeds

At certain times in their production cycle, grass products cannot provide ewes with enough energy or protein. Peak demands are in late pregnancy and early lactation.

To correct shortages, sheep are fed a daily ration which includes energy feeds such as cereal grains (barley and oats), sugar beet feed and by-products with added protein and minerals. These 'concentrates' are high in energy but relatively low in fibre so must be fed with roughage.

Ready-mixed concentrates which are pelleted (compounds) and have a range of energy and protein levels are sold by agricultural merchants and the compounders (Fig 37). They are available in 25kg sacks, and for the small flock this is generally the best way to buy them. Compounders always declare the analysis and usually declare ingredients of a feed. Buyers should ask for the ME figure and discuss the ingredients. Typical ewe compounds have an ME of around 12.5MJ/kg DM and CP 16–18 per cent.

The ingredients in purchased compounds can vary between batches. Because the sudden introduction of a new ingredient may upset the digestion of a ewe in late pregnancy a new ration or new batch should be bought before the previous one is finished so that the old and new can be fed as a mix for a few days.

Feed safety is an important issue in the food chain, so make sure that the supplier can show evidence of raw material traceability and food safety procedures – the customer for lamb may insist on this information.

Organic producers who feed bought-in concentrates will need to source ones which are grown organically, although under some regulations it is permissible for a proportion of the total feed to be non-organic.

Fig 37 Ewes eating compounds from a home-made trough constructed from fibreglass pipe.

Concentrates

Farms which grow cereals can mix their own concentrates and a good basic 14 per cent crude protein concentrate is:

Whole barley	825kg
Soya Bean Meal	150kg
Min/Vit supplement	25kg

This blend will have an energy content of 12.5MJ/kgDM.

For a 16 per cent crude protein concentrate with high UDP:

Whole barley	800kg
Soya Bean Meal	125kg
Fishmeal	50kg
Min/Vit supplement	25kg

This blend will have an energy content of 12.4MJ/kgDM.

These rations can be simplified by replacing the protein and mineral fractions with a single mineralized high protein supplement available from most agricultural merchants.

Cereals are usually fed whole (uncrushed) so that the fibrous husk slows the digestive process and reduces the risk of acidity in the rumen (acidosis). A proportion of the cereal can be substituted by sugar beet shreds or pellets which have four times the fibre content of barley, are popular with sheep, and reduce the risk of acidosis. Trials have shown an improvement in the performance of ewes that are fed sugar beet feed in addition to cereals. Dried grass pellets are another alternative.

Rations must be thoroughly mixed, otherwise sheep can select the bits they like and have an unbalanced diet.

FEEDING CONCENTRATES

Concentrates are fed as a supplement to forage and not as a substitute. They must be introduced into the diet slowly, otherwise the rumen microbes cannot adapt, their activity is depressed and the ewe will stop eating. In late pregnancy this risks pregnancy toxaemia, weak lambs and poor colostrum and milk supplies,

Concentrates are usually fed in troughs (Fig 38) and should be given routinely and as early and late as possible each day. Between feeds the troughs should, where possible, be turned upside down to reduce fouling, especially by birds. Calibrate a bucket so that the rations can be measured accurately at every feed.

An advantage of trough feeding is that ewes can be inspected and caught if there is a problem, and ewes which are off their feed can be identified. But trough space must be adequate to eliminate competition and reduce bullying, otherwise some ewes can eat up to five times as much as others

and the weaker, older and more timid ewes go short of feed.

The disadvantages of trough feeding are that greedy ewes barge and risk abortion when they are heavily pregnant, and it encourages gorging and choking.

Other Feed Systems

Concentrates are also fed in the form of self-help blocks and liquids which give 24-hour access to feed, without regular daily feeding, and allow little and often feeding in contrast to twice daily gorging.

These systems can cost more than the pelleted compounds but some advantages are as follows:

- Convenient to feed.
- Reduce barging and gorging.
- Less prone to loss from rodents and birds (in store and when feeding).
- Changes in intake are a guide to how much grass they have.
- Little and often feeding is better for the digestion.

Disadvantages are:

- Individual feed intakes are difficult to monitor and control.
- Less opportunity to inspect and catch sheep.

Blocks (Fig 39) are usually cereal-based and contain a full range of energy, protein and minerals. They are a convenient way to supplement nutrients in grassland and also improve the digestibility of grass and hay.

Liquid feed is usually derived from molasses and by-products of the distilling industry and is fed from simple ball feeders (Fig 40). The ball revolves and coats itself in liquid as the ewes lick it. Two small (20 litre) feeders are adequate for thirty ewes. Liquids are better than blocks for old ewes with few teeth but the protein source is more often an RDP (urea) than a UDP. The sugars in

Fig 38 Split tractor tyres make cheap, robust feeders.

Fig 39 (Below) *Hill ewes eating feed blocks. Putting blocks on a shallow tray is better than leaving them on the plastic wrappers which make dangerous litter.*

Fig 40 (Below) *Ball feeders, made from 5-gallon containers, dispense liquid feed. The ball system (right) can be bought separately or complete feeders are available.*

molasses aid microbial digestion in the rumen and some liquids may be poured over hay or straw bales to improve their nutritive value.

Feed troughs, blocks and liquids should be rotated around the grazing area to encourage the flock to graze widely and to avoid poaching and a build-up of disease around the feeding area.

Feeding Roughage

Sheep get most of their roughage from grass or from conserved grass such as hay and silage. On arable farms the straw from barley and pea crops is a good source.

Conserved roughage is fed to housed ewes or to supplement pasture. Hay is convenient to feed to small flocks, but silage is suitable. Sheep need about 1.5kg of good hay or 3.5kg of grass silage per head per day for maintenance.

Good silage may be too high in energy for ewes in early pregnancy and make them fat, so it should be rationed.

Spring barley gives the best cereal straw and can be bought in standard bales for

69

easy handling. Ewes take a while to adapt to eating it so it should be introduced early in pregnancy and should not be preceded by hay or silage. About 1.5–2kg per day should be offered fresh, letting them pick out the best and using the rest as bedding. From mid-pregnancy onwards it should be balanced with a high-energy concentrate of at least 12.5 ME and include high levels of a digestible fibre such as sugar beet feed plus 18 per cent CP. Beware of mineral deficiencies such as calcium on this system.

RATIONING THE FEED

Sheep derive at least 90 per cent of their nutrients from forage and in a spring lambing lowland flock a ewe may need only 50kg of compounds and 50kg of hay or 300kg of silage as supplementary feed.

The rule of thumb is that any concentrate feeding should start six to eight weeks before lambing and be introduced at 150g per ewe per day and increased in 100g weekly increments to reach 750–900g per day two weeks before lambing. Around 1kg per day is probably a safe maximum for most sheep.

Feeding should continue after lambing until there is adequate grass or forage (*see* Chapter 6). To avoid overfeeding the rumen, the daily ration should be split into two feeds (morning and evening); when the total exceeds 700g per day it may be advisable to split it into three feeds.

To ensure that the feed is targeted accurately, ewes are usually grouped and fed according to their lambing date; these groups may also be subdivided into groups carrying twins and those carrying singles. Where ewes have not been pregnancy tested, assume that in flocks with a lambing percentage of more than 150 most of the ewes will have twins and in flocks below 150 assume that most will have singles. Then monitor their condition (Table 6) and feed thinner ewes extra, and fat ewes less.

Ewes carrying single lambs may cope satisfactorily on good forage only. Extra feeding is not only expensive but may result in large lambs and dystokia.

Ewe lambs should have a lower daily ration of concentrates than their adult counterparts (around 600g as a maximum) and it should be introduced and fed over a longer period – starting ten weeks before lambing.

At this stage ewe lambs may be changing their teeth (Appendix I, Fig 111) and this can affect their ability to eat root crops or silage.

Targeting the Feed

The key to successful supplementary feeding is to target it accurately at peak production periods.

There are seven periods in the ewe's nutritional calendar.

- The dry period from weaning to flushing.
- Flushing.
- Tupping and the first month of pregnancy.
- Mid pregnancy.
- Late pregnancy.
- Early lactation.
- Late lactation to weaning.

The dry period can last up to ninety days. After weaning ewes can scavenge short pastures while their milk dries up but beware that they are not forced to eat poisonous plants from the hedges. Ewe nutrition from dry, through to flushing, tupping and the first month of pregnancy, is covered in Chapter 4.

In mid-pregnancy the foetus is only 15 per cent of its final birth weight but the placenta – which nourishes the foetus – makes 90 per cent of its growth and is vulnerable to under-nutrition. The level of feeding during this period has a greater effect on the rapidly growing placenta than on the slow-growing foetus. Ewes at condition score 3.5 can be maintained on grass or on a diet giving an ME of 8MJ/day and used to tidy up winter pastures, but bad weather can increase maintenance demands by 50 per cent. A loss of 0.5 in condition score can be tolerated by ewes (but not by ewe lambs) but it is better practice not to deplete fat reserves. If ewes are to be housed this is a good time to introduce them to hay.

In late pregnancy there is fast foetal growth – putting on some 70 per cent of birth weight in the last six weeks. This is when both energy and protein demands are at their highest to grow the foetus and the mammary tissue. Inadequate feeding now can result in pregnancy toxaemia, small lambs, high lamb losses, poor colostrum and milk supply, and poor mothering instincts.

The uterus – filled with the foetus, placenta and fluids – is competing with the rumen for space, and the rumen is the loser. The uterus can fill 60 per cent of the abdominal space, so the ewe is restricted physically in the amount she can eat. She will need a highly concentrated energy source such as a compound of ME 12.5MJ/kgDM and CP 16–18 per cent.

In the last three weeks of pregnancy high-quality concentrates that include UDP should be fed. A ewe supporting multiple foetuses cannot eat enough energy to meet her demands at this stage and must mobilize her own body fat to produce energy; the UDP supplies protein for foetal and metabolic development.

Ewes in late pregnancy have a natural loss of appetite and can quickly go off their feed, so it is important not to over-feed concentrates – a temptation when a heavily pregnant ewe looks a bit thin. Nor should low-quality roughage be balanced by over-feeding concentrates; improve the roughage instead. This is the time to feed the best hay or silage. If straw is the only forage, a compound that includes significant levels of digestible fibre (such as sugar beet feed) is necessary.

The key to feeding ewes in late pregnancy is:

- Increase feed steadily.
- Avoid over and under feeding.
- Improve quality of roughage.
- Try to maintain CS 3.5.
- Do not allow any to lose more than CS 0.5.

Early lactation is the time when ewes seriously metabolize their own body fat, because of the huge energy demands of lactation, and lose condition. In very productive sheep this phenomenon of 'milking off the back' is almost impossible to avoid and is acceptable.

Continue feeding the pre-lambing ration for three to four weeks, by which time the ewe's milk production will have begun to decline. Then, assuming the flock is on grass, roots or good forage, steadily reduce the daily ration of concentrates and discontinue after six weeks. Where there is adequate spring grass (7cm) supplementary feeding can be stopped earlier.

Nutrition for lambs is covered later in Chapter 10.

Table 6 Guide to Condition Score in Ewes			
	Hill ewes	Lowland ewes	
		Outdoors	Housed
8 weeks before tupping	2.5–3.0	3.0–3.5	
Mid-pregnancy	2.5	3.0	
Late pregnancy	2.5–3.0	3.5–4.0	3.0–3.5

This is a general guide. There are variations between breeds.

6 Grassland and Crops

Grassland is the major source of food for sheep. Grazing is the cheapest way to use it but summer surpluses can be conserved as hay or silage for feeding in the winter. One hectare of good grassland can support, throughout the year, ten ewes and their lambs producing around 300kg of lamb carcass or around seventeen finished lambs.

The best pastures for grazing are those with the most leaf and least stem and those containing clover (Fig 41). The most leaf grows in the spring and early summer before the grass has gone to seed. The metabolizable energy (ME) of spring pasture can be 10.6 declining to 9.6 by late summer and to 8 during the winter. Winter grass is leafy because it does not seed, but it is not necessarily nutritious.

Measuring Grass

Grass survives by reproducing and once it has flowered (headed) and set seed the leafy growth declines. Management should discourage seeding by regular grazing and cutting to keep the sward at an average height of 4–6cm.

A simple management aid is to measure grass height with a ruler (Fig 42). The height of green leaf (ignoring stems and seedheads) is measured at forty random points in the field. The height indicates when stock should be moved (under 4cm), when a field is ready to graze (6cm), or when suckling ewes can have compound feed reduced (7cm). (*See* Chapter 5, Table 5.)

Fig 41 Good grassland with a mixture of ryegrass and clover. To get the full benefit of nitrogen fixation and feed value, some 40 per cent of pasture, when viewed from above, should be clover.

Fig 42 To measure grass height, stand a ruler lightly on the ground, slide a finger down until it touches a green leaf – ignoring stems and seedheads – and read off the height.

Target sward heights for grazing are:

Spring
lambing	4–5cm to promote milk.
early lactation	6–7cm to reduce compound feeding.
Summer	6cm to prevent seeding and keep leafy.
Autumn	8cm for flushing; less risk of seeding.
Late autumn/ early winter	3–4cm to avoid winter burn.

Controlling Grass Growth

Average grassland production is around 10–15t DM/hectare/year, but sometimes more than half has grown by late spring; Fig 43 shows a typical growth curve. Grassland should be managed to flatten or utilize the peaks and to extend the grazing season. Ways to do this include:

- Resting the grazing area during the winter.
- Frequent grazing early in the season.
- Choosing an appropriate grazing system.
- Sowing a mix of grasses with different growth patterns.
- Inclusion of clover – which peaks in mid-summer and remains digestible longer in the autumn.
- Conserve peaks as hay or silage.
- Bring in extra stock to graze surpluses.
- Put on fertilizers to encourage growth when it is wanted.

The grazing area should be rested during late winter. Grazing in January and February will reduce growth in the spring by 20 per cent, so any grass needed for spring lambing should not be grazed after Christmas. The flock can graze fields intended for hay until 8–10 weeks before the crop is cut (*see* Hay later).

Frequent grazing reduces the total quantity of pasture but improves the quality and encourages clover. The target height of 4–6cm should prevent overgrazing – which leaves bare patches for weeds to colonize – and undergrazing which allows the grass to go to seed. As soon as grassland grows ahead of the flock it should be topped off mechanically with a mower (Fig 44) or closed off for conservation as hay or silage.

The closer the pasture is grazed the longer it takes to re-grow. In spring it can be grazed close because it grows quickly, but as summer progresses it should not be grazed to less than 4cm. In early winter, grass that will be rested until the spring

73

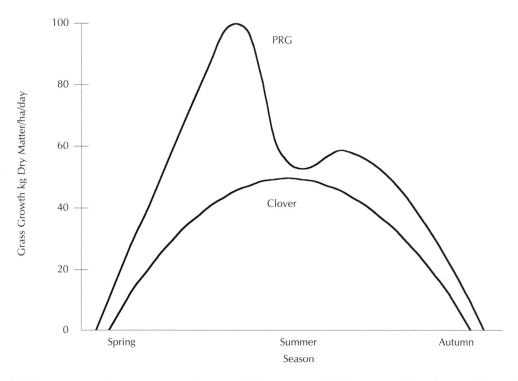

Fig 43 The seasonal production of perennial ryegrass (PRG) and white clover. Most of the grass growth is in the first half of the grazing season. The peak can be flattened by regular grazing (every 14–21 days) and by growing varieties which have a range of heading dates (the date when the ear of the grass emerges) or can be removed for hay or silage. Clover growth peaks in the summer and complements the growth pattern of grass.

can be grazed hard to remove dead vegetation plus green leaf that would otherwise be burned by winter winds. After grazing it can be chain harrowed to spread dung, and to remove debris and decayed grass.

Grazing Systems

There are several systems for grazing sheep. The one selected will depend on many factors but, invariably, it will be a compromise between:

- Set stocking.
- Rotational grazing.
- Strip grazing.
- Forward creep grazing.
- Mixed grazing.

Sheep that are set stocked graze one area for the whole season. Any fertilizer is put on a different third of the area for three consecutive weeks – preferably when rain is imminent. Under this system, judging the quantity of grass and adjusting the stocking rate are difficult.

Rotational grazing involves a number of similar sized areas that the flock grazes in rotation (Fig 45). As each area is vacated it can be topdressed with fertilizer. Grass height indicates when to move the flock and when to close areas for conservation. Sheep can get restless on a system like this that conditions them to moving regularly.

Strip grazing is used in an intensive grassland system and where root and other forage crops are being grazed. Electric fencing is

Fig 44 Expensive machinery is unnecessary for maintaining good grassland. The main jobs are topping off excess growth with a mower (1) and topdressing with a fertilizer spreader (2). Both can be done by a contractor but secondhand mowers and fertilizer spreaders are available, although most need to be powered by a small tractor. A roller (3) and chain harrows (5) are beneficial but the feet and the mouths of a flock of sheep can do the same job. A modern version of the chain harrow is the tine harrow (4). Drainage can be improved by busting hard pans in the soil with a subsoiler (6) but this is a job for the contractor.

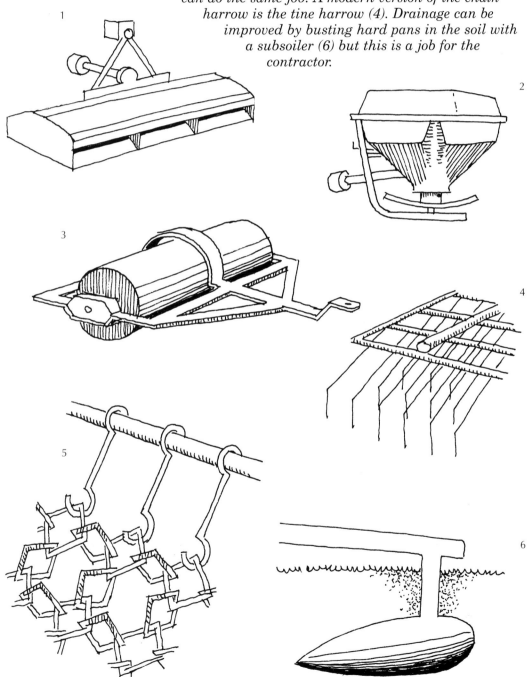

erected across a field to ration the grass and is moved each day to introduce a new strip. The grass can be grazed lightly to give the flock the best, or grazed hard to control it. Sheep get very restless on this system and it is labour intensive.

Forward creep grazing (Fig 45) gives lambs the pick of the grass ahead of the ewes. They access the next grazing area through a creep gate with vertical bars which admit lambs but not ewes. The gate must be sturdy otherwise the ewes destroy it. The system is intended to improve lamb growth rates but results tend to be mixed, and it can cause restlessness and create mud around the gateway. Single lambs are said to creep graze more readily than twins; twins share their milk supply and are reluctant to stray far from their mother in case they miss an opportunity to suck.

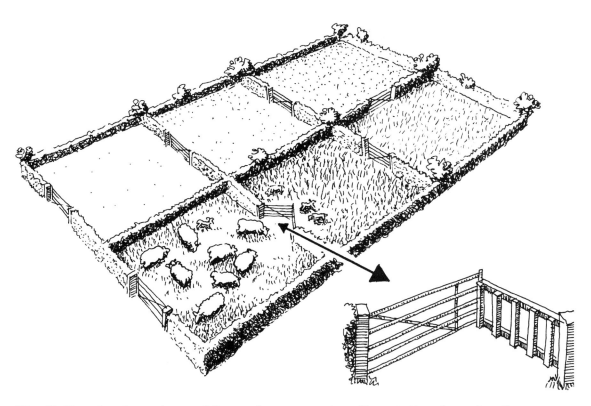

Fig 45 Rotational grazing and forward creep grazing. For rotational grazing the area can be divided into six areas, each one large enough to graze the flock for about four days – thus a complete rotation will take about 24 days. If appropriate the last area to be grazed may be topdressed and topped off. Where grass growth is excessive, one area can be closed for conservation and the cycle changed to five days per field. Where areas are adjacent, forward creep grazing can be practised. Creep gates can be put in the fence or incorporated in a partly opened gate. About 25–30cm gaps in the creep gate will be sufficient for young lambs of most breeds but may need to be widened as they grow. Some of the small primitive breeds or rare breeds may need smaller gaps to prevent ewes getting through. Creep gates must be strongly built and anchored in order to withstand determined ewes.

Mixed grazing is practised under most grazing systems and conventionally means grazing beef cattle and sheep, and also horses. Cattle graze by tearing with their tongues and prefer long grass whereas sheep graze by biting and prefer short grass. Therefore the two species complement each other and utilize more grass. They also graze each other's soiled grass.

Grazing Behaviour

Sheep graze 8–9 hours a day and up to a maximum of thirteen when feed is short. Grazing seems to be controlled by light and is concentrated around the first four hours after dawn and the last four hours around sunset. The implication is that if sheep are yarded at night they should be brought in after their evening session and let out before their dawn session.

On open grazing, sheep will base their grazing on the nearest watering point. On hills they make tracks around the hill because they reach up to graze. On very steep land they make tracks straight up and graze on either side. On flat areas the characteristic waviness in their tracks is due to the sheep watching behind at regular intervals because of the blind spot there is behind their head.

SOIL FERTILITY

The important nutrients for grassland include nitrogen (N), phosphorus (P), potash (K) plus magnesium, sulphur and sodium. The soil should be analysed every three or four years so as to correct any deficiencies or to avoid an over-use of fertilizers which is uneconomic and can cause an imbalance of nutrients. Fertilizer companies and agricultural advisors arrange soil analyses and recommend fertilizer policies. On grazed land most nutrients are recycled and only modest dressings are needed to replace nutrients removed in meat and

wool; where grass is cut for conservation larger replacement dressings are needed.

Soil nutrient status is generally described by an index of 0–3; the lower the figure the less nutrient there is. Acidity is described in terms of pH – the higher the pH figure the less acid the soil. Very acid soil is pH 4.5 and neutral is 7.0. Most grassland needs a pH of around 5.8–6.

Liming

Soil acidity is corrected by lime. Ground limestone and other liming agents are fairly cheap and soil pH tests are made by companies that sell and spread lime. The product is put directly onto grassland or on cultivated soil at any time of the year but preferably on a windless day (Fig 46). As well as reducing acidity, lime will improve soil structure and biological activity and increase the availability of certain plant nutrients. Overliming, on the other hand, can cause mineral imbalances and deficiencies.

Some sources of lime, such as calcified seaweed, are popular with small farmers because they can be applied by fertilizer spreader but these products do not correct acidity as quickly as limestone and are best used annually to maintain pH. Yorkshire fog, lawn daisies, dandelion, moss, plantain, mayweed and sorrel in grassland tend to indicate acidity.

Nitrogen

Nitrogen has a dramatic effect on grass. It behaves like a tap – turning grass growth on and off and generating lush green leaf. An early application in the spring encourages grass for ewes at turnout but excessive use leads to sappy plant growth and scouring.

Nitrogen comes from four sources:

- Soil.
- Clovers.
- Recycled in dung and urine.
- Fertilizers.

77

Fig 46 Liming land to raise the pH of the soil. It should be done on a windless day otherwise the neighbours get the benefit.

Soil reserves are usually highest in land that is grazed.

Clover roots grow nodules containing bacteria that 'fix' atmospheric nitrogen for use by the clover plant. As the old roots die some of this nitrogen becomes available to grass plants. Clover can fix up to 160kg of nitrogen per hectare.

Sheep recycle about 90 per cent of the nitrogen they consume – most of it in the urine and about 30 per cent in the dung.

Nitrogen is usually bought in 50kg bags of a nitrogenous fertilizer, most commonly in the form of ammonium nitrate.

Grass begins to grow when the daytime air temperature reaches around 10°C or the soil temperature is 5.5°C at 10cm. The first nitrogen fertilizer can go on two weeks before this, but because nitrogen is wasted on grassland which is not growing, a more accurate way to time the first application is the Dutch T-sum method:

- Record the average daily air temperature in centigrade from 1 January.
- Ignore minus figures.
- When the figures add up to 200, nitrogen can be applied.

Some farming magazines publish the weekly T-sum for each region. However, the decision to apply nitrogen should also take into account:

- The soil conditions (no good if waterlogged or damaged by tractor tyres).
- The weather forecast (no good if it is going to turn cold).
- When the grass is needed (no point in growing it too early).

Nitrogen is usually applied little and often on grazing land. The bulk is spread early to promote early growth and late to extend the grazing season for flushing. However, applied nitrogen like any fertilizer is wasted unless the grass it produces is converted into sheep meat, milk or conserved feed.

Phosphate and Potash

Nitrogen is applied to give a quick response and any residue at the end of the growing season is wasted. Phosphate and potash, on the other hand, are applied to maintain levels in the soil from year to year.

Both are necessary for root development and plant growth and can be applied with nitrogen in compound fertilizers that are formulated to provide a balance of nutrients (*see* box). The timing of phosphate and potash applications is less critical than for nitrogen and may be applied at any time. However, potash must not be topdressed in the spring on land that is being grazed by lactating ewes because it can reduce the availability of magnesium and trigger hypomagnesaemia.

Much of the phosphate and potash is recycled by the grazing animal – phosphate in the dung and potash in the urine – but lambs, hay and silage remove considerable amounts that need replacing. Around one hundred standard bales of hay can remove 85kg of potash.

Organic Fertilizers

Organic fertilizers tend to release nutrients slowly to sustain growth rather than produce dramatic growth. Because the nitrogen is less soluble than in inorganic fertilizers it is less easily leached (dissolved and removed)

Using Fertilizers

The nutrient requirements of soil are expressed in terms of kg/hectare. The nutrient contents of fertilizers are expressed in percentages. It is necessary to reconcile the two in order to know how much fertilizer to buy.

A common grassland fertilizer is expressed as a '21:8:11'. The figures represent N, P and K content respectively, and their percentages. So this fertilizer contains 21 per cent nitrogen, 8 per cent phosphate and 11 per cent potash. If the bag is a standard 50kg bag the amount of soil nutrients in each bag are:

N (21 per cent of 50kg) = 10.5kg
P (8 per cent of 50kg) = 4.0kg
K (11 per cent of 50kg) = 5.5kg

This means that putting it on at the rate of three bags per hectare gives the grassland 31.5kg of nitrogen, 12kg of phosphate and 16.5kg of potash.

An average clover/grass pasture with average P and K reserves for a spring lambing flock carrying about ten ewes per hectare might need total annual dressings of 100–120kg N, 36kg P and 36kg K, per hectare respectively.

A basic programme for grassland could be:

Early spring 25–30kg/ha N for early growth (Some P could be applied now)
Late spring 25–30kg/ha N and 12kg P. N sustains growth. Grass responds to P in the spring because only the surface roots are active and they cannot seek out soil phosphate.
Early summer 25–30kg N, 12kg P, 18kg K. K is needed on conservation areas. N to sustain growth.
Late summer 25–30kg N to extend grazing for flushing and finishing lambs. 12kg P, 18kg K to top up soil reserves.

High nitrogen applications are not recommended for hay because the lush grass is difficult to dry, but where extra N is used it should be matched by the same amount of extra K.

by rain. Most provide extra minerals and trace elements, and promote bacteria and earthworm activity.

GRASSES AND CLOVERS

Grazing pasture contains a mixture of grasses and clovers (Fig 47) which can survive persistent grazing. They do this by having a prostrate growth habit (Fig 48) to protect the growing points from being eaten. They should also have a wide range of heading dates (date when the ear or seed head emerges) so that leaf production is spread throughout the season. Grassland for conservation, on the other hand, has a narrow range of varieties and heading dates because the crop is harvested at one time. Owners of small flocks will probably compromise and use a seed mixture that is suitable for both grazing and cutting.

Perennial ryegrass (PRG) is an important grass species and has a wide range of heading dates. It forms a dense sward, survives hard grazing and is suited to long-term pastures (*see* later). Italian ryegrass, on the other hand, has a shorter life, early growth and is best suited to early grazing and conservation in short-term pastures, usually associated with dairy cattle.

Fig 47 Common species used in productive pastures.

1. *Timothy*
   (Phleum pratense*)*
2. *Cocksfoot*
   (Dactylis glomerata*)*
3. *Perennial ryegrass*
   (Lolium perenne*)*
4. *White clover*
   (*Trifolium repens*).*

Fig 48 The growth habit of ryegrass. The prostrate varieties (above) are better for grazing; their growing points are less prone to being eaten than in the erect varieties (below) which are better for conservation. Seed mixtures can include a range of ryegrasses to extend the grazing season and to allow some conservation.

Timothy is winter-hardy and grows well in cool, wet areas; Cocksfoot has a strong root system and thrives on light or droughty soils; Meadow fescue tolerates a range of climatic conditions and low fertility.

Seed merchants are a good source of advice on grassland, and their sales leaflets carry useful information. They design grassland mixtures to suit the locality, the soil, the purpose, the grazing systems, and the level of fertilizer inputs.

Clover

White clover is palatable, survives persistent grazing, is high in energy, protein, minerals and vitamins and it fixes nitrogen. Trials show that lambs gain weight 20 per cent faster on a clover sward than on pure grass, but sheep must have gradual access to swards with a high clover content

to prevent digestive upsets. It contributes up to 160kg of nitrogen per hectare and has a higher and more stable feed value than grass, but is more temperature-sensitive, grows later in the spring and is difficult to manage and maintain.

Clover should comprise about 40 per cent of the sward when viewed from above. This can sustain ten ewes per hectare without extra nitrogen but for higher stocking rates, or to grow grass for conservation, the best compromise is to apply an annual total of 120kg per hectare in four equal applications during the season. High levels of nitrogen fertilizer (200kg per hectare per year) reduce clover by encouraging excessive grass to shade it; grassland should be managed so as to keep clover exposed to sunlight.

Seed merchants formulate mixtures for high clover pasture. Those containing 20 per cent clover seed can give 30–50 per cent clover growth in the pasture, peaking in midsummer and sustainable for at least five years. Only one light dressing of nitrogen is

---

Getting the Best from Grassland

- Keep it leafy by grazing, conservation, topping or adjusting stocking rate.
- Keep at the optimum height.
- Encourage productive grass species and clover.
- Avoid overgrazing and poaching.
- Keep out weed infestations by maintaining a dense sward with no bare patches.
- Have regular soil analyses.
- Keep soil nutrients and pH at appropriate level.
- Improve drainage if necessary.
- Adopt a grazing system that suits the sheep and benefits grass management.
- Make best use of the seasonal surpluses.
- Extend the grazing season at both ends.
- Do not graze too late in the winter.
- Keep field records.

needed in early spring to get the grass growing before the clover is active. The mixtures contain small and large leafed varieties; the large are more productive, but the small withstand grazing better and give good ground cover to prevent ingress of weeds. Wild white clover is especially important for sheep because it tolerates their very close grazing and fills gaps in the swards. The two genuine varieties are Kent and S184.

Red clover is less tolerant of grazing and its high oestrogen content reduces ewe fertility, although small quantities are unlikely to be a risk. On acid soils, where clover will not thrive, trefoil – another member of the clover family – is finding a niche. It tolerates soils of pH 5 and less, fixes nitrogen and survives lax grazing. It needs a non-aggressive grass companion. Making a comeback is sainfoin, a highly palatable forage legume (clover is a member of the legume family) with a high protein content that produces quality grazing and hay. In a similar category is lucerne or alfalfa.

Types of Pasture

Grassland is usually categorized by life expectancy. Permanent pasture is normally found on an all-grass farm and is grassland that is not in an arable rotation. With a significant perennial ryegrass content it is productive, can resist poaching and drought and is cheap to maintain, but needs good management to stay productive.

Long-term leys are expected to last for more than five years; medium-term for three to five years and short-term for one to three years. Short-term leys are used in rotation with other crops and are usually based on Italian ryegrass that gives early growth but are more suited to conservation than to persistent grazing.

Because ploughing and reseeding is expensive, long-term and permanent pastures have a cost advantage, but pastures that are in a rotation allow weed, disease and pest cycles to be broken.

IMPROVING GRASSLAND

Grassland which is not very productive can be improved by:

- Management techniques.
- Over-seeding.
- Reseeding.

Neglected and under-grazed grassland (Fig 49) that contains around 50 per cent ryegrass is improved by topping – to remove grass stems and weeds such as thistles and nettles – and by harrowing (Fig 44) to remove matted grass and debris at ground level. Nutrient deficiencies and drainage problems should be corrected. Persistent weeds such as bracken, docks and nettles, and creeping grasses and thistles that spread by underground or prostrate stems (rhizomes and stolons), will need regular topping or even spraying with a herbicide.

After this, hard grazing will encourage ryegrass to tiller (produce new shoots), become dominant and replace unproductive grasses. Initially it is worth erecting an

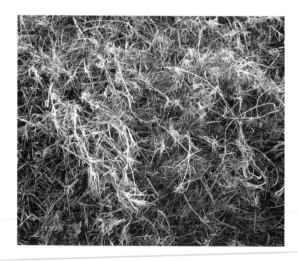

Fig 49 Unproductive grassland with unpalatable dead grass. An ideal candidate for hard grazing then over-seeding with ryegrasses and white clover to improve its productivity.

electric fence, borrowing a herd of cattle and persuading them to graze hard – without poaching the ground.

Over-seeding

Pastures which have less than 50 per cent ryegrass and are dominated by 'weed' grasses can be over-seeded. So-called weed grasses, such as Yorkshire fog (*Holcus lanatus*), annual meadow grass (*Poa annua*), Crested dogstail (*Cynosurus cristatus*), common bent (*Agrostis tenuis*) and soft brome (*Bromus mollis*), are low yielding, have a high stem to leaf ratio, can be very invasive and are unresponsive to nitrogen; on the other hand they are useful on poor soils. Palatable weeds such as dandelions and young docks can provide variety and valuable minerals in the diet.

Small areas can be improved by the 'scatter and tread' technique. Scatter a vigorous perennial ryegrass mixture on the moist surface, preferably in late spring or autumn, and let the flock tread it and graze for two to three days. The technique is a gamble because the seeds need quick treading followed by two weeks of continuing moisture – but not too wet – to allow them to establish.

This system is also used to increase the clover content of a pasture – sowing small-leaved white clover.

Over-seeding is also useful where weeds have been killed, leaving bare patches. Nettles and thistles shade productive grasses, take space and are unpalatable. They can be rogued (pulled by hand), or spot sprayed with a knapsack sprayer or watering can. Observe withdrawal periods for livestock on any treated areas and wear the recommended protective clothing when spraying.

Larger areas, which are difficult to cultivate or cannot be taken out of production, are over-seeded using machinery. Early autumn and spring are suitable times; autumn-sown clover can be hit or miss, but it may go dormant and then germinate when the conditions are right.

Correct the soil pH and topdress with 12kg N, 60kg P and 60kg K per hectare and drill or broadcast the seeds. Bare soil is chain harrowed and the seeds that are broadcast are rolled but those that are drilled may not benefit from rolling because it closes the slot and delays germination. Continue grazing to tread the seed.

Over-seeding is best with quick growing species such as tetraploid ryegrasses but include slower growing species in the mixture. Aim for a sowing rate of 22kg per hectare or two-thirds of the full reseed rate.

Pasture can also be renovated by sod seeding or direct drilling. This is usually done in the autumn by a specialist contractor who drills seeds direct into the pasture after it has been grazed short and perhaps partially desiccated by chemicals such as glyphosate.

Reseeding

New swards are established in spring or autumn because summer drought or winter cold will kill the crop if it is sown outside these periods. Old pasture is best ploughed in the autumn and left to break down over winter, but to destroy weeds and to break the cycle of disease it is useful to grow a break crop such as kale or rape (see later) before reseeding with grass.

Soil acidity should be pH 6 and the potash and phosphate levels should be corrected by topdressing the seedbed with around 60kg per hectare of each (depending on the soil analysis). The final seedbed must be fine and firm and the seeds can be broadcast with the fertilizer from a fertilizer spreader, or drilled. Graze lightly and quickly as soon as there is some growth – after six to ten weeks, depending on the season – to encourage tillering, to bite off annual weeds and to consolidate the ground.

Drainage

Poor drainage is no good for either grassland or sheep; neither will thrive when their feet are permanently wet. Minor waterlogging may be due to soil compaction by regular treading. Ploughing to the same depth for many years creates a hard pan under the soil surface that impedes the drainage of surface water and reduces root development. A subsoiler (Fig 44) at a depth of about 300mm busts the pan and creates fissures with minimal disturbance on the surface. A specialist drainage company can advise on drainage problems.

CONSERVING GRASS

Excess growth in the summer is cut and saved as hay (dried grass) or silage (pickled grass) for feeding in the winter. This also keeps grassland under control, and the aftermaths offer fresh grazing for finishing lambs or for flushing ewes.

Hay is easy to store and feed and one hectare can provide enough for a thirty ewe flock for the year.

Sheep are taken off the grassland that is intended for hay about eight to ten weeks before the crop is cut. The best time to make hay is during the longest days, so in the northern hemisphere the grass will be 'closed' for hay in early April and cut during the third week in June.

The grass is cut when it begins to flower – that is, when it begins to shed pollen from the seed heads. If it is cut earlier the quality of the hay may be higher but the yield will be lower. It then needs five consecutive days of fine dry weather during which time it is turned two or three times to dry it evenly. It is baled when it feels dry, rustles and is easily broken (20 per cent moisture). After baling and storing the hay continues to cure and if it is damp it heats up; excessive heating will cause a fire.

It is usually best to employ a contractor to do the job although a flock owner with a small tractor might reduce costs by investing in a hay turner and letting the contractor cut and bale the hay.

Haymaking is weather dependent, very time consuming and unpredictable. Unless organic hay is wanted it is often easier to shop around and buy it. Normally it is cheaper to buy it straight out of the hay field but make sure that it is in the standard small bales; large bales are not very convenient. Beware of any that is made late in the season because it may contain weed seeds, thistles and poisonous ragwort.

The best hay has a high clover and leaf content, minimum stem and a green tinge. It should not be dusty or smell mouldy. Hay that has been made during a spell of wet weather will have had nutrients washed away and may be dark and dusty. This dust may contain spores of *Micropolyspora faeni* that, as well as affecting sheep, will cause a crippling condition in humans called Farmer's Lung.

Sheep eat about 1.5kg per head per day of hay (plus some waste) and although flocks that graze winter forage crops may need very little, allow ten bales per head per year to provide for a hard winter, or for a dry summer when hay may have to be fed to supplement grass. Hay stores quite well and it is wise to carry some over from each year. Standard bales weigh 20–30kg (33–50 bales per tonne) and large bales weigh around 500–600kg. An average crop of grass should produce around 250 standard bales/hectare.

Silage

Silage is grass that is pickled in acid produced naturally by fermentation in anaerobic conditions. Grass should be cut on a sunny dry day and wilted for twenty-four to forty-eight hours to reduce the moisture content. It is made either by compacting the wilted grass in a clamp and excluding the air or compressing it into big bales that are made air-tight by being wrapped in polythene. Again, this is a job for a contractor. Quality silage has an ME of 10.5 and sheep eat it readily but it has been associated with listeriosis, probably due to soil contamination.

For the small flock on an all-sheep farm, silage is difficult to justify because it is awkward to handle, and once a store is open it deteriorates quickly. Silage is fed at around 3.5kg per ewe per day so a thirty ewe flock would take several days to eat a 0.5 tonne big bale which, once open, may become inedible before it is eaten.

Owners of very small flocks ensile surplus grass in plastic bin bags. It should be made in the summer when it is warm and the sugar levels (important to fermentation) are high. Mow with a lawnmower then wilt until no moisture can be squeezed from it; pack into airtight dustbin bags, squeeze the air out of the bags and seal the ends with a tie or fold over and use a castrating ring and applicator.

OTHER CROPS FOR SHEEP

Root and forage crops such as swedes, fodder beet, stubble turnips, rape and kale complement grassland. They not only feed ewes and lambs during lean periods but provide a break in the grass to control weeds and disease. Forage crops such as stubble or Dutch turnips, forage rape and kale are called 'catch' crops and give cheap forage within ten weeks of sowing. They are broadcast at any time from spring to late summer to provide grazing until Christmas, or the land can be regrassed in the autumn. One hectare can give six to eight weeks' grazing for sixty lambs but there is considerable waste due to trampling unless it is strip grazed.

The crops are grazed *in situ* but a 'runback' area to grass must be provided to give a clean, dry lying area and to allow sheep to be introduced gradually onto the crop. They are best suited to free draining soils. Hay should be provided in racks to ensure that sheep have sufficient dry matter intake on these low dry matter feeds.

Root crops such as swedes and fodder beet (Fig 50) provide winter and spring feed because they can be lifted and stored. A hectare of swedes can feed sixty lambs for eighteen weeks and 8kg replaces 1kg of barley. Fodder cabbages provide excellent winter feed and are either grazed *in situ* or cut and carried.

These crops occupy the whole growing season. Most are drilled in late spring, fed during the winter and the land prepared for reseeding in the following spring.

Forage rye or Westerwolds one-year ryegrass are crops that are sown in autumn in worn-out grassland, to give very early grazing in the spring, and then followed by reseeding or by an arable crop such as barley.

A local farmer may be willing to grow a forage crop on his land; it would provide a break in his rotation, improve soil fertility and give him a cash crop. Agree a price beforehand; the yield cannot be guaranteed so paying a sum per head per week of grazing ensures that you get what you pay for.

Fodder Beet

Fodder beet is a perfect crop for the small flock, although it requires some experience to grow. A close relation of sugar beet and mangold it is succulent, palatable, high in energy, high yielding and can replace some or all of the concentrate ration for pregnant ewes and store lambs. It is also popular with other farm livestock.

Sown in late spring in a good tilth with a soil pH of 6.5 they are usually lifted in early winter before being frosted, and will store until the following spring. Sowing, weeding, lifting and storing can either be done by hand or can be fully mechanized.

At lifting the tops should be trimmed off to prevent deterioration in the store; tops can be fed to sheep but should be wilted for four days to reduce the oxalic acid content. The roots store well in a straw-covered clamp protected from the frost.

Fodder beet can be fed whole on the ground but young ewes, that are changing teeth, and old ewes with loose teeth may need them roughly chopped. Avoid soil contamination and feed them with hay, silage or straw at about 60 per cent of the total daily dry matter intake otherwise they may cause scouring.

Fodder beet will give a total yield of around 90t per hectare so, fed at 5kg per head per day, 0.25 hectares will feed a thirty ewe flock for more than twenty weeks.

The crop has relatively few disease problems but can be affected by eelworm.

Agroforestry

Agroforestry is the integration of trees and farming. Sheep integrate well with trees, despite their tendency to browse, and are often grazed in orchards. Trees for agroforestry are planted at 5m or 10m intervals and protected by 1.5m plastic tubes; this

Fig 50 Fodder beet is an excellent crop for the small flock. High yielding and nutritious, it is a cross between the mangold and sugar beet. It lends itself to full mechanization or to hand cultivation and can be stored and fed from early winter to spring.

spacing allows enough light to maintain good grassland. The sheep control the weeds and reduce the need for herbicides or regular cutting; the trees provide income and shelter, and are environmentally attractive. In Denmark, sheep are integrated with Christmas tree plantations. They are introduced when the trees are two years old and the key to success is to avoid over-stocking or letting them run short of grazing.

Some breeds have a greater tendency to nibble trees, especially if stressed, and the Danes have identified the Shropshire as being compatible with forestry.

7 Wool and Shearing—

The world produces annually more than three million tonnes of greasy wool but, on many sheep farms, wool takes second place to meat production. The development of man-made fibres and the importance of sheep meat has made wool a by-product. Nevertheless, owners of small flocks will find it worthwhile to make the most of wool by:

- Managing and feeding to encourage wool growth.
- Selecting stock for fleece weight and quality.
- Breeding to eliminate kemp (*see* later), hair and unwanted coloured wool.
- Taking advantage of fleece assessment services.
- Protecting against skin and wool diseases.
- Keeping fleeces clean.
- Using a competent shearer.

Fig 51 Wool grows in staples and is crimped. Fibres should be even in length and thickness, and without hair or kemp.

- Careful storing to avoid damp.
- Good presentation of fleeces at marketing.
- Adding value through private sales or manufacturing.

Wool is a remarkable, renewable fibre. It is so elastic that it can be stretched repeatedly to up to a third of its normal length and still go back to its original length; this elasticity is the reason that wool garments hold their shape. It can absorb up to 30 per cent of its own weight in moisture but at the same time can repel liquid.

The fleece grows in small tufts (staples) of fibres (Fig 51) and each fibre is wavy (crimped) to give bulk and springiness to the wool. Wool is an excellent insulator because it traps air by this combination of crimp and fine fibre.

There are basically three types of fibre in the adult fleece – wool, hair and kemp. True wool fibres are the finest, kemp fibres are short and coarse and are shed seasonally, and hair fibres are intermediate in size. Fig 52 illustrates the differences between wool and hair. Hair is not wanted in a fleece because it causes off-dyeing and will flatten under pressure.

All wool fibres are produced from structures in the skin called follicles (Fig 52) of which there are two types. The primary follicles grow the coarser wool and hair. They appear early in the life of a foetus and at birth are fully developed and are growing fibres. After birth no more follicles are formed, so the nutrition of the pregnant ewe has an important influence on the future wool production of her lamb.

The secondary follicles produce the finer wool. These are not fully developed at birth and continue developing during the first

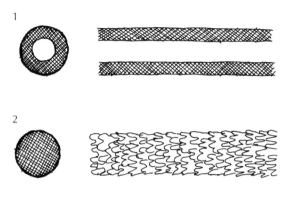

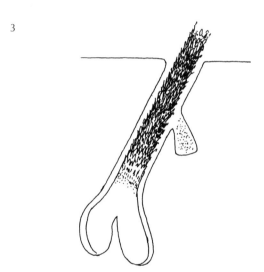

Fig 52 *A section of hair (1) shows that it has a largely hollow centre and a smooth surface. Hairs are a problem in fleeces because they fail to dye properly and will collapse under pressure. Wool fibre (2) has a more solid centre and the surface is covered with scales much like roof tiles. Wool fibres are produced by follicles (3) in the skin. Primary follicles grow the coarser wool and secondary follicles, which develop later, grow finer wool. No more follicles are formed after birth but secondary follicles continue to develop.*

few months of life. This is why some fine-wool breeds are born almost bald and susceptible to hypothermia. They also take longer to complete fleece development. A coarse-wool Romney, for example, might complete follicle development at around one month of age but a very fine-wool Merino lamb could be five months old before all the follicles have developed.

The implication of this is that at birth some lambs have fleeces that are more coarse (stronger) and less woolly than their final fleece. Therefore breeding stock should not be selected for fleece weight and quality until it is at least a year old when the fleece has fully developed.

WOOL GROWTH

Many sheep produce wool in excess of what they need for warmth. Wool grows continuously; even when an animal is starving it will grow wool at the expense of other organs and may continue to grow wool after death. Wool growth is very sensitive to nutrition and the wool follicle responds quickly to a shortage of feed by reducing the diameter of the growing fibres (Fig 53).

After a temporary food shortage there is a narrowing of each fibre at the same level throughout the fleece, resulting in a weak area or 'break'. This thinning weakens the wool and affects the manufacturing potential, downgrading the fleece. Some producers shear at around lambing time to ensure that a break caused by the stress of pregnancy will be at the bottom of the fibres and not in the centre where it would reduce the value. Breaks often occur in ewes which have been stressed, but if the problem is widespread in the flock it is possible to judge from the position of the break in the fibre when it occurred and, hence, what might have caused it, to prevent it in future. 'Reading' breaks is one way to monitor flock management.

Wool growth is seasonal and seems to be affected by daylight hours. In British hill

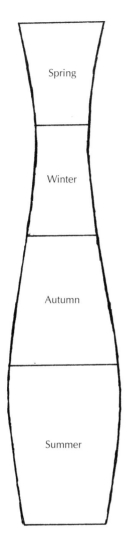

Fig 53 Typical growth pattern of a wool fibre showing seasonal variations in length and diameter. The fibre narrows at times of stress – usually in the winter and during pregnancy – and can result in a severe weakness or break in the wool.

breeds, winter growth may be as little as 30 per cent of summer growth.

Age also influences fleece growth; it increases to its maximum at four to five years and then declines – sometimes rapidly. An increase in energy – as long as protein levels are sufficient – can increase wool growth in just a few days.

The back of the sheep is the area of highest wool production and anything damaging the fleece on the back – such as rubbing or weathering – reduces weight.

WOOL QUALITY

Wool is sold by weight and the prices reflect the quality of the wool and the current market demand. Generally a heavy fleece will be worth more than a light fleece, but quality is important for two reasons:

- Quality attracts a premium price for the producer.
- Quality makes manufacturing easier and encourages a wider use of wool.

Quality is judged from a sample cut from the shoulder or side – the areas that are representative of the whole fleece. The features to look for are:

Good quality factors:

- Springiness. Spring is due to crimp. A tight, uniform, well-defined crimp gives resilience, easier spinning and garments keep their shape.
- Softness. Is usually associated with fineness and is desirable.
- Staple length – should be even.
- Colour. Unless coloured wool is wanted for specialist use, white is essential for dyeing a uniform colour.

Poor quality factors:

- Breaks. Stretch the wool to check for weak areas.
- Kemp and coloured fibres. These restrict the dyeing process.
- Cotting. This is when the fibres become matted like felt.
- Discoloration. Caused by skin disease such as mycotic dermatitis and by weathering.

• Contamination. The whole fleece should be clean.

Managing for Quality

Man-made faults are the scourge of the wool industry and are usually the result of carelessness.

Contamination with synthetic string, excessive marking fluid, dags, earth and moiety (vegetable matter such as straw and hay) are common. In addition, cigarettes, tools, coffee mugs and bits of clothing have all been found in wool sacks.

Moiety is difficult to avoid and is worse in housed sheep that are fed hay in overhead hay racks, and in sheep outwintered on muddy root crops. In an extreme system of wool production Merino flocks are housed, protected by coats and fed a specialist diet to produce perfect, clean, fine wool.

Damp fleeces deteriorate, so sheep should not be shorn when wet and fleeces must be stored off the ground and away from walls.

Keep identification sprays and paints to a minimum and always use scourable brands. Avoid coloured dips used to tint sheep for shows and sales. Cotting is found on old ewes and those that have been ill, but it can be reduced by improved nutrition.

Disease

Discoloured and damaged wool due to parasites and disease should be controlled by veterinary treatments. Skin diseases spoil the fleece, usually because of excessive itching and rubbing, and cause the animal considerable distress. Lice, for example, can reduce fleece weight by up to six per cent.

Mycotic dermatitis, which is rife in wet areas, shows up at shearing as a yellow crust and stains on the skin and at the base of the wool – usually along the back. In small flocks individual sheep can be treated by clipping the affected areas and rubbing in powdered alum – usually available from pharmacies.

Most skin problems, such as sheep scab, lice, ticks and fly strike are preventable and are covered in Chapter 8. Skins have a value when stock are sold for slaughter and should not be allowed to be damaged by disease, shearing, injections or barbed wire.

Breeding for Quality

Most characteristics of wool are highly heritable, so there is an opportunity to select breeding stock to:

• Raise fleece weights.
• Increase the fineness.
• Eliminate or reduce coloured fibres and kemp.

Wool comes primarily from the ewe flock so attention to fleeces when selecting ewe lamb replacements is important. But a change of ram will make the quickest progress. The British Wool Marketing Board and other organizations offer 'wool on the hoof' and ram fleece assessment services, and the serious breeder can use the services of specialist laboratories to measure a number of fleece characteristics to assist in selection. Ram breeders may provide a fleece assessment certificate with the animals they sell.

Fleece weight is moderately heritable in most breeds and through selection can be increased by 1.5 per cent a year; it takes eight generations to get a 12.6 per cent increase. By weighing the fleeces at shearing and then calculating a mean fleece weight for the flock or for individual age groups, individual sheep can be ranked as above or below the average for their group.

To produce fine wool, rams from breeds such as the Merino, Corriedale and Polwarth make quick improvements when used over most breeds of ewe. Generally, sires of a breed with a white face produce crossbred offspring with very white fleeces.

Grey fibres and kemp can be bred out quickly by using rams that do not have these faults.

Irregular quality and length are often due to poor selection and too much mixed blood. First crosses give good wool, but mongrels are rarely satisfactory.

WOOL MARKETING

In the UK the British Wool Marketing Board has a statutory obligation to market all the wool grown in the country, and all new producers should contact the Board for registration. This system is a boon to the majority of growers, but producers of specialist wools, such as very fine or coloured wools, may find more lucrative outlets for their fleeces. They will need to agree this with the Board.

Once a new producer has signed on with the Board they will be allocated a local haulier and wool depot. The haulier will supply wool sacks, collect the wool and take it to the depot. Producers can visit the depot, see their wool being graded and have any faults explained.

Fleeces that are sold direct to hand spinners must be uniform in length, clean, attractively presented and labelled with the weight and breed. Spinners prefer fresh fleeces and if they are stored they must be kept cool and dry. Some mills will spin small numbers of fleeces for customers to sell or to use in a cottage industry making woollen goods.

The Markets for Wool

Fibre length and fineness dictate the uses for wools and, therefore, the market.

Hill breeds such as the Scottish Blackface and Welsh Mountain have long strong fleeces of around 2kg, suited to making carpets, rugs and blankets. Longwools such as the Leicester, Wensleydale, Lincoln and Devon and Cornwall Longwool, found on grassy lowlands, are big sheep with long lustrous fleeces of 10kg suited to speciality fabrics.

Down breeds such as the Dorset, Hampshire and Suffolk have close, fine, short wool of 2–3kg that is good for knitting and for soft woollens.

Coloured breeds such as the Jacob and Black Welsh Mountain are popular with home spinners and the Merino, Southdown, Shetland and Charollais have some of the finer fleeces.

SHEARING THE FLOCK

Flocks are usually shorn annually in early summer for husbandry reasons as well as to harvest the wool. In some countries twice yearly shearing is practised, resulting in about 10 per cent more wool.

Shearing is easiest when the 'rise' has taken place. The rise is the point in the fleece where the previous season's growth has finished. Above that point the fibres may be matted, and below it the new growth is clean and straight and easier to shear.

Pre-lambing or winter shearing of housed ewes is common. Shearing ewes at housing avoids contamination from straw and hay and can result in heavier birth weights (up by 0.5kg), better survival rates and higher weaning weights. Less space is needed (stocking rate can be increased by 20 per cent) and the ewes are less heat stressed.

Other advantages are that winter shearing is at a quiet time in the shepherd's year, ewe body condition and the lambing process are more visible and lambs find it easier to find a teat. At turnout the shorn ewes tend to seek shelter and take their lambs with them. However, shorn ewes eat more and may produce large lambs which cause dystokia.

Housing should be draught-free with plenty of bedding and the flock should be shorn at least eight weeks before turnout using a standard comb to leave 5mm of fleece; there should be 12mm by turnout and plenty of shelter. Restrict the practice to fit spring-lambing lowland ewes, and avoiding older or thin ewes.

Shearing lambs in the autumn encourages growth and keeps them clean during the winter.

Shearing is traditionally done by hand but chemical and robot systems have been developed in Australia. Chemical shearing produces a weak zone in the fibre which can then be peeled off by hand. Experimental work resulted in wool being lost in the field before harvesting and it left no protective stubble. Robot shearing has had some success but is not widely practised.

Electric shearing equipment is expensive, so the small flock owner could learn to shear with blades (Fig 54). Courses in all types of shearing are usually run by colleges and the British Wool Marketing Board. However, professional shearers are readily available. They normally charge a flat rate to shear a small flock and make a charge per head to shear large flocks.

Shearing should be planned well in advance because there is likely to be a lot of last-minute indecision especially when there is risk of rain. Professional shearers need to be able to shear at a moment's notice. The shearing area should be ready and clean and with sufficient wool sacks available.

The aim of good shearing is to:

- Minimize stress on the sheep.
- Produce a clean fleece.
- Let the shearer work efficiently.

To minimize stress keep sheep off food or on bare pasture (with water) for at least twelve hours beforehand. Full stomachs during shearing can cause discomfort or even death.

Before shearing, and preferably before housing or penning, pick off any contamination, such as vegetable matter, and crutch any sheep which have dung or urine-soaked wool. The sheep must be dry; they can be held under cover prior to shearing but must be dry when brought under cover and must not be bedded on sawdust or straw which will contaminate the fleeces. Slatted floors or green bedding such as long grass and docks are suitable but bare concrete produces a slurry soup which ruins fleeces.

Ask the shearer well in advance what facilities and labour he wants. He may have

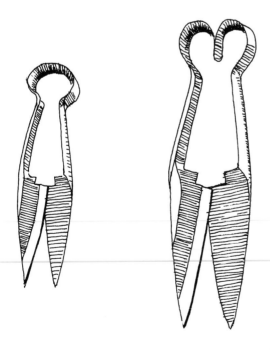

Fig 54 Blade shearers can shear more than 300 sheep in an 8-hour day. For the small flock a pair of shears is essential for dagging, and for trimming around wounds, mycotic dermatitis and fly strike. There are different designs for shearing, dagging and trimming for showing. Two common ones are the single bow (left) and the double bow. The double bow requires a lighter squeeze than the single bow. Recommended blade lengths are 13.5cm for dagging and 16cm for shearing or trimming. A small pocket version (left) is handy for everyday husbandry use. Trimming show sheep is best done with a special bent shear. The blades should be kept dry, and wiped with an oily cloth after use. When they are correctly sharpened, they should 'sing' and cut a hair or piece of paper cleanly.

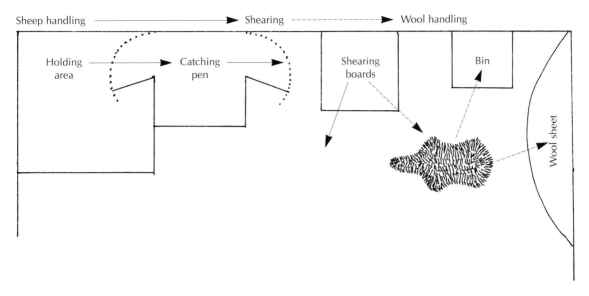

Fig 55 A simple shearing layout must ensure a flow of sheep and a flow of fleeces. The shearing boards and the fleece rolling areas must be kept clean at all times. The wool sheet should be hung for easy access and a bin will hold any dags and contaminated wool. Lamb's wool comes off the lamb in loose bits and should be packed in separate sheets.

a shearing trailer complete with catching pens and a shearing platform. In this case he will need only a holding pen from which to fill his catching pens, and a power supply.

If he does not have a trailer he will need a well lit, sheltered working area with a clean, level floor. The shearing area should be about 2m × 2m for each shearer, close to the catching pen and have a hook from which to hang the shearing machine – probably from a roof beam or a vertical wooden post. Have a secure holding pen (Fig 55) which can be reduced in size as the flock shrinks and an adjacent catching pen for about ten sheep.

Check in advance that the power – which may be a secondary supply to an outside building – is safe and dependable and does not blow trip switches or contact breakers.

He will appreciate a tea break during or after shearing and somewhere to wash. The basic job requirements are:

- Shearing.
- Catching: to catch and hand sheep to shearer (if there are no self-closing doors on the pen to enable him to catch them himself).
- Keeping the shearing boards swept of dung and bits of fleece.
- Keeping the catching pens full.
- Throwing, tidying, rolling and packing fleeces.

With planning, these jobs can be done comfortably by one shearer and two assistants.

Handling the Fleece

Once the fleece is shorn there is an art in 'throwing' it to land in the wool packing area the correct way up – and with the neck at one end and the crutch at the other. Unless it falls correctly it cannot be rolled correctly. The shearer may throw it himself or teach the flock owner.

The area for rolling and packing must be clean – beware of any hay or straw which may be stored nearby.

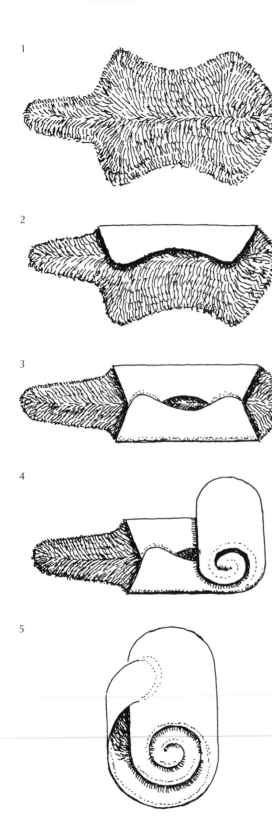

Fig 56 In the northern hemisphere fleeces are usually rolled and packed in wool sacks or sheets. In the southern hemisphere they are skirted (wool is taken from the edges to reduce variation in the fleece) and pressed into bales. The rolling method consists of (1) throwing the fleece onto a flat, clean surface usually flesh side down; (2 and 3) folding in the flanks then (4) rolling firmly from tail to head and (5) tucking the neck securely into the fleece.

Check fleeces again for dirty wool and vegetable matter and then roll (Fig 56) and pack tightly into the sacks or sheets keeping breeds separate, white and coloured fleeces separate and fleeces from different classes of sheep – ewes, rams, hoggets and lambs – separate.

Shearing Tips

- If a sheep goes into shock during shearing, throw a bucket of cold water over its head.
- Keep a marker handy to identify a sheep that has its ear tag shorn off.
- Have a broom and constantly keep the shearing and rolling areas clean.
- Warn the shearer of any problems, such as a vaccination abscess, and treat any cuts with an antiseptic spray. Small cuts are common and are healed quickly by the lanolin in the skin. Serious cuts are not acceptable.
- Shear coloured sheep last and keep fleeces separate from white fleeces.
- Have a spring balance and identify, weigh and record individual fleeces for a breeding programme.
- Beware of wounds attracting flies. Treat with a repellent or cream.
- A sheep's appetite increases by 25 per cent after shearing, so ensure extra grass.

8 Keeping the Flock Healthy

Cynics say that a sheep's only ambition in life is to die. In fact sheep are remarkably tough animals and, other than during an outbreak of disease or attacks by predators, adult sheep losses should not average more than about 4 per cent a year. Most of these losses occur around lambing time.

The job of the flock owner is to:

- Keep a flock healthy.
- Recognize a sick sheep.
- Diagnose the problem – a combination of experience, a veterinary surgeon and a veterinary book (*see* Further Reading).
- Nurse a sick sheep back to health.

The essentials for maintaining health are:

- Adequate nutrition.
- Minimum stress.
- Safe environment.
- Observation.
- Prompt diagnosis and treatment.
- Good nursing.
- Preventive medicine.
- Hygiene.
- Good flock management.

Nutrition

Nutrition is the single most important factor in maintaining a healthy flock. Loss of appetite indicates ill health but check that the sheep is able to eat. Look for sharp, broken or bleeding teeth or a wad of cud compacted inside the cheek. In these cases the ewe will probably need preferential feeding or culling.

Stress

Many health problems are triggered by stress. Stress is anything that puts the body under strain, such as pregnancy, inadequate nutrition, extremes of weather, rough handling and being chased by dogs. Many life-threatening organisms live harmlessly in the body and only cause disease when body defences are lowered by stress.

Observations have shown that weather affects health. In a wet, cloudy year with low sunshine hours, the rain reduces the feed value of herbage and the lack of sunshine affects general health. The results are often poor ewe condition, low lambing percentages, slow lamb finishing and reduced wool weight.

Environment

A farm is a dangerous place and sheep have ingenious ways of injuring themselves. Some common accidents include:

- Breaking limbs or getting a head stuck in netting fences.
- Getting entangled by the fleece in hedges, brambles and barbed wire.
- Pulling bale stacks over.
- Being stuck (cast) on the back.
- Cuts and injuries from fighting – especially rams.
- Injuries from bolting through hedges and barbed fences.
- Trapping a hoof or puncturing the sole with a thorn or nail.
- Hanging in a loop of baler twine.
- Falling in rivers and ditches.

Predators such as dogs are a danger. Stress takes its toll and sheep need calming after an attack. Dog-worrying in the UK and losses from coyotes in the USA have encouraged the use of guard dogs – popularly the Komondor, Maremma, Pyrenean and Akbash. Other guard animals that run with the flock include the donkey, that apparently dislikes dogs and kicks them, and the llama and ostrich. Anyone resorting to this method of predator control should check their insurance and be aware that some of these guard animals may fail to differentiate between predator and friend. The law is usually on the side of the flock owner when it comes to shooting predators, but the regulations should be confirmed.

Poisoning damages or destroys vital body tissues and organs. It can be caused by grass contaminated with a weedkiller, an excess of a medicine, an excess of copper, or when forced to eat poisonous plants such as rhododendron, yew or ragwort during a drought or feed shortage. Sheep find dried ragwort palatable, so ragwort in hay is a risk.

Sheep enjoy chewing, especially something toxic like lead painted woodwork or unripe acorns. They choke on string, small potatoes and thistle heads, and on compound feeds eaten too quickly. Choking sheep usually foam at the mouth, cough and walk backwards but generally manage to dislodge the object.

Sheep become cast when skin parasites are an irritation or in showery weather when drying skin causes itching. They roll over to rub their backs on the ground and cannot get up. Ewes in full wool and heavily pregnant are most at risk. Controlling skin parasites (see later) can reduce the problem as does a back scratcher such as an old trailer. Regular checking is essential as cast sheep rarely bleat and a heavily pregnant ewe can die quickly. Secondary danger comes from carrion birds which attack the soft tissue such as eyes and crutch. Ewes which have been on their backs for some time will be bloated and stagger when helped to their feet, but will recover. Often it is the same ewes which regularly become cast.

Observation

A change in behaviour is the first sign of a health problem. So it is necessary to:

- Recognize normal behaviour.
- Observe the flock regularly.
- Know the sheep.
- Know what problems to expect, and when.

Recognizing normal behaviour comes with experience. It is important to see changes of behaviour in the context of the production cycle. A ewe that isolates herself from the flock during lambing time is likely to be having a lamb. At any other time of the year she is likely to be ill. Looking at the flock regularly – ideally twice a day in lowland flocks – means that changes in behaviour can be spotted early.

Knowing individual sheep is an aid. If a normally fat sheep becomes thin, or a normally nervy ewe becomes passive, suspect a problem.

Problems can be spotted quickly if the shepherd anticipates them. He should know the common diseases and the time of year or the conditions that favour them.

For example:

- Pregnancy toxaemia before lambing.
- Mastitis after lambing and weaning.
- Fly strike or head fly during the fly season.
- Pneumonia in humid weather and after handling or stressing.
- Worm burdens in young grazing lambs.

Classic changes in behaviour include:

- Staying apart from the flock.
- Recumbent or reluctant to move.
- Persistent rubbing and biting at flanks.
- Not eating or cudding.

- Grinding its teeth.
- Abnormally nervy or abnormally placid.

Classic physical changes include:

- Abnormal faeces (normal faeces are firm and pelleted) such as diarrhoea.
- Abnormal posture or gait.
- Eyes dull or crusted.
- Nose discharging.
- Salivation.
- Ears drooping.
- Swellings or sores.
- Wool loss.
- Shivering or convulsions.
- Breathing laboured or shallow – normal respiration rate ten to twenty per minute.
- Temperature above or below the normal of 39.0–40.0°C. Put a veterinary thermometer about 5cm into the rectum for up to two minutes.

Diagnosis

When a shepherd sees a problem he does not recognize or cannot cope with, such as a difficult lambing, he should involve a veterinary surgeon or perhaps initially a local experienced shepherd who is willing to help.

Provided it is fit to travel, and the surgery is within reasonable distance, it is better to take the animal to the surgery where there are facilities.

However, in a surgery the veterinary surgeon cannot see the animal's natural behaviour or environment, so it is essential to offer information such as:

- Age and breed.
- Stage in production cycle.
- Any history of illness – the individual and the flock.
- Symptoms.
- When they were first noticed.
- Any other similar cases in the flock.
- Recent treatments such as worming, vaccination, dipping.
- Recent changes in feed or pasture.

- Access to poisonous plants or other poisons.
- Routine nutrition such as free access minerals.
- Changes in environment – being housed, moved, weather.
- Stress such as transporting, being dogged, yarded, weaned.

When visiting the veterinary surgeon take the opportunity to ask questions and, after treatment, it is helpful to him to let him know the outcome.

A veterinary surgeon is responsible for the drugs he prescribes and may want to see an animal or visit the flock before giving a prescription. But once he has a good relationship with a client, and trusts their judgement, he may be willing to supply some drugs without seeing the animal. Always involve a veterinary surgeon as soon as possible; the longer a problem is left, the less the chance of success.

Nursing

If the veterinary surgeon is the doctor then the shepherd is the nurse.

When a treated animal is returned to the flock it must be easy to catch for further treatment or inspection. Otherwise a sick sheep should be bedded on straw in a well ventilated, well-lit, draught-free area that can be disinfected.

A recumbent sheep should be turned regularly to prevent sores and discomfort, and preferably propped up with bales to prevent bloat. Sheep need to be able to keep their head up for belching.

Water (warm and slightly salty) and food must be accessible – a recumbent sheep cannot drink from a bucket on the other side of a pen.

As soon as an animal shows signs of wanting to stand it is important to encourage it by lifting (not by the wool) and supporting it for short periods in an improvised sling or between bales. Sheep should be lifted onto their hind legs first because that is the natural way for them to stand up.

An animal off its food should be tempted to eat. Antibiotics may kill bacteria in the rumen which will upset the digestive system and depress appetite. There are veterinary products that put microbes back into the rumen to get it working.

Persuading a sick ewe to eat is a labour of love. Tasty morsels include molassed sugar beet feed and coarse molassed cereals but titbits out of the hedgerows, such as dandelions and ivy leaves, can bring a sheep back from the brink. It may be necessary to push a wad into the mouth to encourage chewing. Success has been reported with a dose of Guinness, with a serving of Shredded Wheat with powdered glucose, and with a mixture of a little table salt with black treacle.

A sheep that stops drinking must be given water or an electrolyte (commercially available mixtures of dextrose, sodium chloride and potassium) using a drenching gun. If the animal is conscious, water and even liquidized feeds can be given by stomach tube (calf-sized) to save time, stress and the risk of choking.

If a sick ewe has lambs these can be kept with her but they will need hand feeding if she cannot suckle them. Whether or not to take a sick lamb away from its mother depends on the situation at the time, but in the long term it is better to keep them together. A very sick ewe is unlikely to need company but once she starts to fret she may be well enough to go back into the flock or may need a companion.

Giving Medicines

Medicines are primarily given orally (Fig 57) or by intramuscular (Fig 58) or subcutaneous (Fig 59) injection. Fig 60 illustrates the injection sites.

Always follow the instructions on medicines and observe the withdrawal periods.

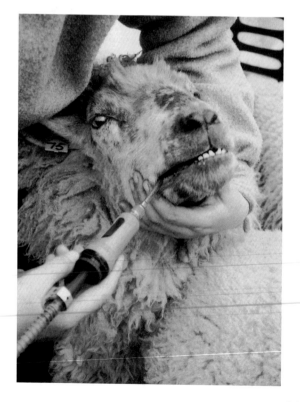

Fig 57 Drenching a ewe with an oral dose of medicine. Drenching guns are calibrated to deliver the precise amount and have a shaped nozzle to prevent injury. Tilt the head up very slightly and put the nozzle through the side of the mouth where there are no teeth and dose over the back of tongue. Oral worm drenches should go into the rumen to be effective. If they go into the mouth rather than over the back of tongue the drug may by-pass the rumen and go direct into the other stomachs. Keep out of the lungs – pneumonia can be triggered by careless drenching.

Work slowly and keep animals well restrained in a race or tightly mobbed in a pen to restrict movement. Lambs, in particular, tend to jump forwards and injure their throats, sometimes with fatal results. For small numbers, or for the occasional dose, use a plastic syringe with the nozzle cut off. A 20ml size is suitable for adults and a 2ml or 5ml for lambs.

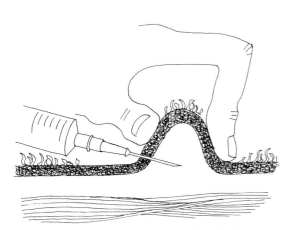

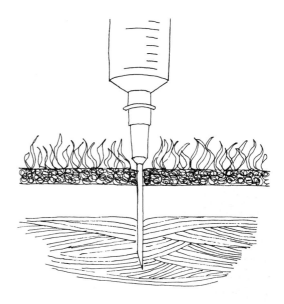

Fig 59 (Left) **Subcutaneous injection**. *This route is usually used for vaccines and large dose medicines such as calcium borogluconate to correct hypocalcaemia. Administer just under a loose area of skin with the needle parallel to the body to avoid the muscle. Beware of entering the fold of skin one side and out the other. Leaking may be a problem, so massage gently to disperse. A large volume should be warmed to blood temperature and divided between several sites. A veterinary surgeon may put some of these medicines directly into a vein, but the layman must not.*

Fig 58 (Right) **Intramuscular injection**. *This route is usually used for antibiotics. The substance is absorbed rapidly and distributed around the body. It should be given on a fleshy site, avoiding bones, and on a different site each day when a course of antibiotics is being given.*

This is the time that must lapse between giving a medicine and having the animal slaughtered for human consumption or drinking its milk. Where a number of treatments are involved it is safer to write out a programme.

It is usually essential that a course of treatment is completed, even if the animal has recovered.

Some medicines are not compatible, so mixtures of drugs, antibiotics or vaccines should not be used without consulting the veterinary surgeon.

Small quantities of injectable medicines are normally supplied in a disposable syringe with disposable needles. They should be stored safely and cleanly according to the instructions – usually in the refrigerator – and disposed of safely.

Most flock owners will need their own non-disposable syringes. The 10m, 20ml and 50ml sizes will cover most uses. Sheep needles are preferably 2.5cm × 19 gauge. For short-term treatments, such as a course of antibiotics, disposable needles are preferable and should not be used for more than six injections.

There are two risks with needles; they blunt and they spread disease. For repeated injections, such as flock vaccinations, a non-disposable needle should be changed after every twelve sheep. Where a product is being drawn from a bottle, keep one sterilized needle in the bottle through which to fill the syringe.

Sterilize non-disposable needles and syringes in boiling water for ten minutes; wrap the needles in linen and aluminium foil

Fig 60 **Injection sites**. As a principle, subcutaneous injections should be made in loose skin areas such as the neck region and intramuscular injections where there is good deep muscle. But the sites should be chosen so as not to enter nerves, veins or bones and not to damage expensive cuts of meat.

Part the wool to get a clean area of skin. Avoid contaminating the needle by dropping it on the fleece. Expel air bubbles from the syringe.

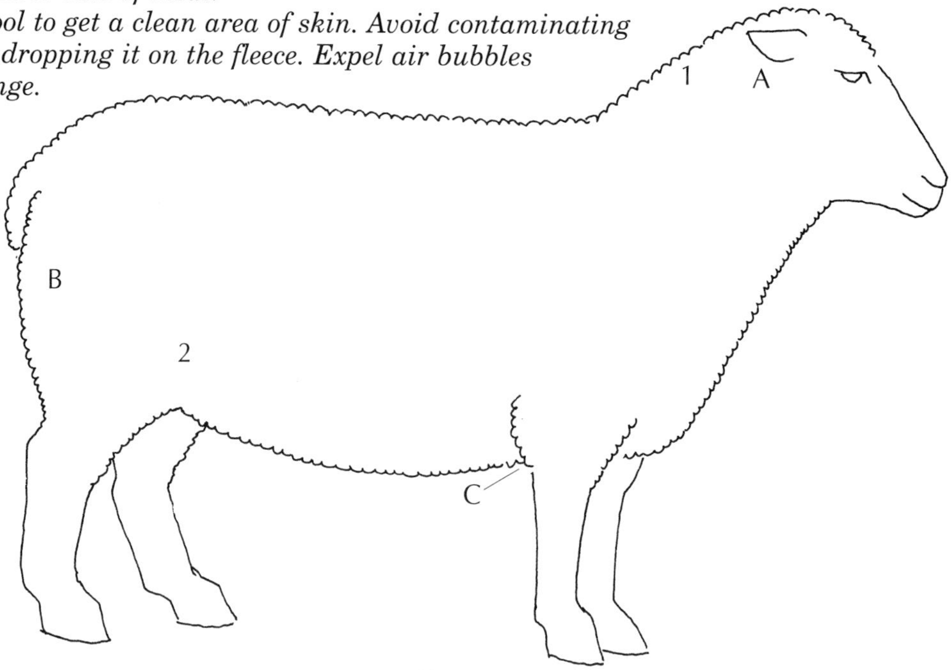

Lambs:
1. Subcutaneous – a fold in the skin in the back of the neck.
2. Intramuscular – in the big muscle at the top of the front of the hind leg. This is safer than in the back of the leg because there is less risk of nerve damage.

Ewes:
A. Subcutaneous – top of the neck behind the ears or in the mid neck region.
B. Intramuscular – hind leg where there is good muscle.
C. For orf, scratch the bare skin inside the top of a front leg.

to prevent blunting and leave them in the package until needed. Chemical sterilants are effective but can affect some drugs.

PREVENTIVE MEDICINE

Preventive medicine is any good husbandry or management that keeps sheep healthy. Health may be protected by vaccinating against specific diseases or simply by keeping fences secure so that neighbouring flocks cannot mix and spread infection.

Preventive medicine includes:

• Health schemes.
• Vaccination programmes.
• Flock management to minimize infection.
• Hygiene.

National flock health schemes operate in most sheep countries to eradicate or control specific diseases and provide a register of accredited flocks. Target diseases include Scrapie, Enzootic Abortion and Maedi Visna. Flocks that are shown to be clear of a disease (usually through blood testing) benefit by improved productivity and by selling breeding stock at a premium. The small flock may profit from a scheme but will need to defend health status with good fencing and care with regard to buying-in replacement stock.

A private flock health scheme can be drawn up with the veterinary surgeon. The box below is an example.

Private Health Scheme

Five visits per annum by the local veterinary surgeon.

Visit 1
Eight weeks before tupping.
- Check ram fertility and condition.
- Condition score a selection of ewes.
- Take blood samples to check for trace element status and abortion agents.
- Faeces sample for worm count.
- Advise on a forage analysis for fibre, protein and energy.

Visit 2
Six weeks before lambing.
- Check condition of flock.
- Discuss feed and nutrition especially trace elements and protein source; treat any clinical cases.
- Advise on vaccination against clostridial disease, worming, etc.

Visit 3
Two weeks before lambing.
- Discuss veterinary needs for lambing and arrange for preventive medicines.
- Advise on or treat any problems.
- Discuss housing and lambing facilities.
- Check condition of ewes.

Visit 4
Peak of lambing.
- Check and advise on any problems.
- Demonstrate how to deal with some problems such as tubing, intraperitoneal injection, dystokia.
- Advise on hygiene and disease control such as *E. coli*, coccidia.
- Discuss worm-control programme.

Visit 5
Twelve weeks after lambing.
- Discuss post lambing problems and check growth rate, mastitis cases, etc.
- Discuss vaccination programme for the year.
- Discuss worm-control programme, including dog tapeworms.
- Discuss footrot control-programme.
- Discuss dipping and external parasite control.
- Review any conditions that arose such as metabolic problems, abortion, pneumonia and mastitis for future control.

Vaccination

Vaccination confers immunity on an animal by introducing specific pathogens – either live or modified – into the body. These stimulate the body to produce antibodies that react with the pathogens and suppress or kill them. Not all vaccines are used the same way; depending on the type of vaccine some are given as a single dose and some as a double dose (primary and secondary) a few weeks apart; some pass immunity to lambs and some do not; some must be used at a specific time in the calendar; some are given as an injection and some as a scratch; some must not be given to pregnant sheep. Some are transmissible to humans and must be handled with care.

Never vaccinate sheep when they are sick, wet or dirty and try to avoid combining vaccinations with too many other activities such as dipping and worm dosing, especially during pregnancy, because the stress can be fatal.

It is essential to read the instructions and write out a flock programme. The veterinary surgeon will advise on suitable vaccinations based on local experience. Top of his list will be a vaccination against a range of clostridial diseases and against pneumonia. Clostridial diseases include tetanus, pulpy kidney and lamb dysentery. The clostridial bacteria are widespread in the environment and kill their victim by manufacturing toxins in the body; they have such good survival mechanisms that they cannot be eradicated. Other vaccines include orf, toxoplasmosis, enzootic abortion (EAE), louping-ill and footrot.

Hygiene

Hygiene is a major factor in preventing disease. Any facilities or equipment that are shared with other flocks – such as shearing, transport, dipping – should be cleaned between use. Needles should be sterilized. Housing and lambing equipment should be disinfected immediately after turnout. Working areas should be cleaned after use – such as clearing away foot parings. Cats should be kept away from feed stores and hay bales because they spread toxoplasmosis – one of the common causes of abortion.

Flock Management

A number of major health problems can be controlled by management. These include:

- Footrot.
- Internal parasites (worms).
- External parasites.

FOOT PROBLEMS

Footrot is an infectious and debilitating disease. It is caused by two bacteria which thrive in damp conditions when the air temperature is above 10°C. Physical signs are limping, grazing on the knees, reddening between the digits (Fig 61), pus in the hoof and a characteristic smell.

Some sheep may be genetically susceptible to the condition so avoid selecting replacements, especially rams, from stock which are persistently lame.

Footrot may be prevented by regular foot care, preferably during dry weather.

1. Check and trim (Fig 62) the feet of each animal.
2. All those with sound feet should be turned onto dry pasture that has been free of sheep for at least fourteen days. The bacteria that cause footrot cannot survive outside the sheep for more than fourteen days, so this pasture should be clear of infection.

Where there is a footbath (Fig 30) these clean sheep should stand in it according to the instructions and then on concrete for two hours before being put on the pasture. Zinc sulphate products are widely used as a footbath chemical because they penetrate the hoof and reach any infection.

3a. Those with suspect feet should have any infection exposed and treated with an antibiotic spray. Cases which do not heal may need an antibiotic injection.

3b. They should be kept separate from the first group.

3c. The group should be re-examined, treated three days later and then weekly until cured.

3d. Cured animals should be footbathed and returned to the 'clean' flock.

Protection can be given by a vaccination used before the high risk periods but it is

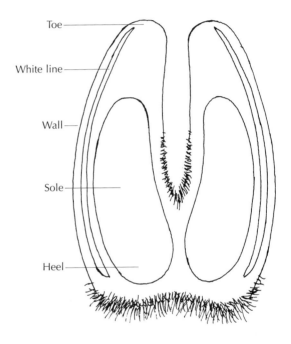

Fig 61 Sole view of a sheep's foot showing the points. Problems occur between the two halves (digits) such as scald, mudballs and stones. Infection gets into the hoof causing the wall to separate from the sole at the white line.

Fig 62 Hoofs need to be inspected and trimmed regularly to prevent them from becoming overgrown. A veterinary surgeon or experienced shepherd will demonstrate the basics of trimming, but the aim is to maintain the natural shape of the foot, to leave enough wall to protect the sole and to prevent or cure any infection by removing overgrown horn. Trim cautiously, especially across the toe, to avoid cutting a blood vessel. Modern hoof shears are powerful and must be used with care. Where footrot is established, paring allows oxygen and the footbath chemical or antibiotic spray to reach the infection.

Over-trimming can leave a strawberry-like growth (granuloma or proudflesh) in the sole which never heals and needs veterinary attention. When the outer wall is loose (shelly hoof) and soil becomes impacted it may cause an abscess; loose wall should be trimmed away.

only effective when used in conjunction with regular trimming and inspection. Once eradicated from the farm footrot can only return with bought-in, borrowed or straying sheep.

Other Causes of Lameness

Not all lameness is caused by footrot and other causes should be suspected. It may simply be mechanical such as a stone or ball of mud between the digits. Or a puncture from a thorn.

An abscess at the top of the hoof is common; it may not be visible but the hoof may feel warm. Where it is visible it should be cleaned and dressed with an antibiotic spray. Untended it can penetrate bones in the hoof and need antibiotic injections to clear. As with many infections, if the sheep is otherwise healthy it will heal easily.

Scald is common in lambs on lush grass that softens and abrades the skin between the digits and allows infection to enter. The earliest sign is a limp but if more than one foot is affected they walk as if they have arthritis. A look between the digits reveals moist pink skin or pus; a squirt of antibiotic spray heals it quickly.

Limping may also be the result of a leg injury rather than a problem with the hoof. It is also associated with mastitis when a ewe swings a rear leg to avoid a painful udder.

INTERNAL PARASITES

Internal parasites, known as worms or helminths, are a major worldwide problem in sheep flocks. They include:

- Stomach and intestinal worms (roundworms or nematodes) that cause Parasitic Gastro-enteritis (PGE).
- Liver fluke that burrow in the liver.
- Tapeworms that cause cysts in the liver and brain (gid and hydatid disease) and are dangerous to humans.

The main causes of Parasitic Gastro-enteritis are the roundworms *Ostertagia circumcinta* that induce weight loss and scouring and *Haemonchus contortus* that suck blood and precipitate anaemia. They are a problem in late summer and autumn.

Nematodirus battus cause nematodirus infection in grazing lambs of one to four months old, most commonly in late spring and early summer. The onset of the disease is sudden, causing acute diarrhoea and dehydration manifest by excessive drinking. Ewes are not affected.

Parasitic Gastro-enteritis may be confused with poor nutrition and trace element deficiencies. Veterinary surgeons can diagnose from dung samples but this is not always a good guide. Response to wormers (anthelmintics) is a means of diagnosis – if a dosed sheep stops scouring within two to three days then worms were likely to have been the problem.

In theory roundworms can be eradicated by breaking their life cycle (Fig 63) either by killing the parasites in the sheep with a wormer or on the pastures with a desiccant. In reality neither are totally effective. Work continues to perfect a vaccine against worms, and to find a chemical or a parasite to kill eggs in the pasture or dung.

However, the cycle does allow the shepherd some control. The strategies are:

- Keep grazing 'clean' from infection
- Dose with wormers at strategic times to reduce infection.

At its simplest, 'clean' grazing is pasture which has not been grazed by sheep for twelve months. Most roundworms cannot survive away from the sheep for that length of time.

Clean grazing may be provided by:

- A three-year rotation (Fig 64) with one year cattle, one year conservation and one year sheep, with lambs moving onto the conservation aftermaths.

104

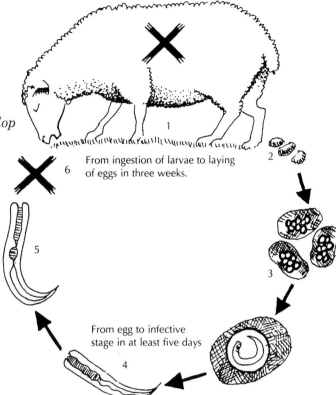

Fig 63 A typical sheep roundworm cycle illustrates how the cycle can be broken in order to keep worm infection under control.

1. *Infective larvae in the sheep develop to adults and lay eggs.*
2. *Eggs voided in the dung.*
3. *Worms develop in the egg in the dung.*
4. *Young worms or larvae hatch.*
5. *Larva develops to infective stage.*
6. *Infective larva climbs up grass and is eaten by the sheep.*
X *Two places where the cycle can be broken.*
   Stop the sheep picking up infection by providing clean grazing.
   Stop the adults laying eggs by worming the sheep.

From ingestion of larvae to laying of eggs in three weeks.

From egg to infective stage in at least five days

- A two-year-rotation with cattle and sheep – exchanging the grazings in the spring and dosing all the stock at changeover. This is not strictly 'clean' because some worms infect both species.
- Hay or silage aftermaths on a sheep-only grass farm. This gives some clean grazing for weaned lambs but the proportion of conserved land to grazing land is small.
- A new ley sown after an arable crop.
- Growing short-term leys and forage crops in a rotation. Forage crops grown immediately after pasture may still be infected.
- Have very low stocking rates to mini-mize pasture contamination.

The reality is that where there are sheep there are worms and in most flocks they are controlled by treating sheep with wormers.

These are administered as a drench, an injection, a bolus or in the feed.

To use wormers effectively it is useful to see what happens on worm-infested pastures. Fig 65 shows the natural pattern of infestation on pasture without control. Eggs deposited in the first half of the grazing season cause dangerous levels of infection later.

The goal of a worm control programme is simply to reduce the level of worms contaminating the pasture. A programme must be tailored to individual flocks, but it can be based on strategic treatments.

Key treatments for ewes are:

1. One dose at flushing – it also improves fertility.
2. One dose either at housing, before lambing, at lambing or before turnout in order to reduce contamination from the

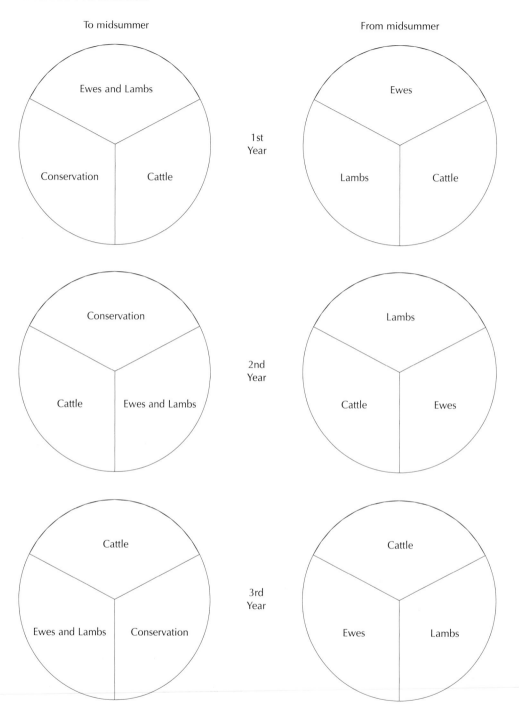

To midsummer

From midsummer

1st Year

Ewes and Lambs

Conservation

Cattle

Ewes

Lambs

Cattle

2nd Year

Conservation

Cattle

Ewes and Lambs

Lambs

Cattle

Ewes

3rd Year

Cattle

Ewes and Lambs

Conservation

Cattle

Ewes

Lambs

Fig 64 A clean grazing system based on dividing the grassland between cattle, sheep and conservation. To get a balance between cattle and sheep the starting point, in grazing terms, is that one adult bovine is equal to five adult sheep. So, in theory, five cattle will need as much grazing as twenty-five ewes. The system obeys the principles that weaned lambs should go onto 'clean' aftermaths, and young lambs should not graze pasture which was grazed by young lambs in the previous two years.

106

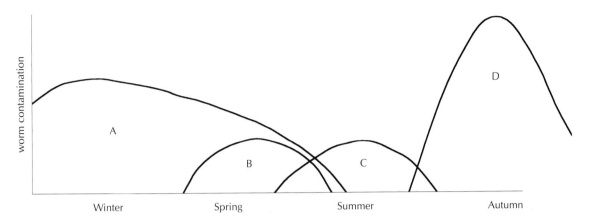

Fig 65 The pattern of natural worm infestation on pasture in the absence of a worm control programme. Eggs deposited in the first half of the grazing season are responsible for dangerous levels of infection in the second half.

A. *Overwintered infestation on pasture. It remains inert until temperatures reach 10°C then becomes infective. Infestation will tail off in early summer as the larvae die.*
B. *Eggs shed by ewes at lambing time.*
C. *Eggs produced by lambs which ingested the over-wintered infestation earlier in the season.*
D. *Massive infestation in late summer caused by the development of worms from B and C.*

The aim of a worming programme is to turn out onto clean pasture to avoid A, drench ewes at lambing to prevent B, drench grazing lambs to prevent C. This programme should prevent D.

'spring rise'. The spring or periparturient rise is the natural increase in worm eggs that are shed by a ewe at around lambing time when hormonal changes lower her ability to suppress worms.

If turnout is onto clean pasture, further dosing may not be necessary.

Where there is low-risk pasture:

• Give one dose six weeks after lambing or via a wormer feed block.

Where there is no clean pasture:

• Dose every three weeks during lactation to reduce contamination on the pasture and to combat infection from over-wintered eggs.

Lambs and Worms

PGE does not normally strike lambs until midsummer and those born before the end of February may miss the risk period. However, lambs grazing contaminated pasture early in the grazing season will act as a host to worms and allow them to multiply. When lambs begin to graze contaminated pasture – at about a month old – they should be dosed thereafter at three-weekly intervals. It takes three weeks from when larvae are ingested until the eggs are laid, so a three-weekly dose prevents the eggs from being laid and causing further contamination.

Always dose before moving to cleaner grazing. If lambs go onto clean aftermaths at weaning they should not need further dosing; if there is no clean grazing at weaning, dose until sold. Lambs being kept as replacements must continue on a worm-control programme because worm infestations leave lasting damage to the gut, lungs and liver. Weaned ewes can stay on contaminated pasture because by now they have built up their immunity.

Nematodirus Infection

Young grazing lambs face an additional threat – from Nematodirus infection. This is spread from lamb crop to lamb crop and the ewe plays a negligible role. The rule is never to graze lambs on land that was grazed by lambs in the previous spring and preferably not for two springs. Land that has only grazed adults should be safe. Although spring is the traditional time for Nematodirus problems, some farms have experienced outbreaks in autumn.

The larval stage of the *N. battus* is unusual in that it stays in the egg and can survive on pasture for up to two years. They normally hatch after a cold spell followed by a mean temperature of at least 10°C. This means that eggs accumulate on the pasture and an explosive outbreak is timed to coincide with the natural start of the lamb grazing season. It also means that hatching can be predicted from the weather conditions. Lambs at risk should be dosed at three-weekly intervals starting just before a predicted outbreak using a wormer which covers *Nematodirus* spp.

Lambs are at risk when:

- Aged one to four months old (after this they develop immunity).
- Climatic conditions are right for hatching.
- Grass was grazed in the previous year by lambs.

The disease can be avoided by:

- Clean grazing.
- Lambing early or late to avoid the risk periods.

USING WORMERS

Some internal parasites are developing resistance to wormers and it is the responsibility of sheep producers to prevent it. In practical terms, once resistance has struck it will be forever.

Wormer resistance is a heritable characteristic, with the offspring of resistant worms carrying the genes. The more frequent the use of wormers the more pressure there is on selection, and resistance can develop quickly. With infrequent worming the rate of selection will be minimal. Underdosing (not giving the full dose for the weight of the animal) also exerts strong selection pressure and leads to rapid development of resistance. Resistance is possible in all parasiticides in the sheep industry so none of them should be either misused or over-used.

If sheep do not respond to a wormer that has been correctly administered, the veterinary surgeon can make a faecal test to check for resistance. If there is resistance, continuing with that chemical will make the problem worse.

Some of the rules for avoiding resistance are:

- Change the wormer chemical each year. Make the change at housing/turnout and do not change more often. Changing the brand of wormer does not necessarily mean that the chemical has been changed. There are only three major groups of chemicals:

  Benzimidazoles (white wormers)
  Levamisoles (clear wormers)
  Avermectins/milbemycins

- Follow the instructions.
- Have a strategy to reduce the number of dosings, such as a clean grazing system.
- Do not keep sheep and goats together. Resistance is common in goats.
- Do not underdose. Weigh each animal before dosing or divide into groups of

similar size, weigh the heaviest in each group and treat all animals in the group at this dose level.

- Check the accuracy of the calibrations on the gun by measuring several doses of the wormer (not water) into a measuring jar or syringe.
- Products which are in suspension need careful shaking, but not too much otherwise air bubbles make the dose inaccurate.
- Do not mix different wormers.
- Avoid unnecessary dosing.
- Avoid introducing resistant worms. Dose bought-in sheep with an avermectin on arrival, because currently there is no known resistance to this group, then keep them off the pasture for twenty-four hours.

Choosing Wormers

Most wormers are 'broad spectrum' – they kill a wide range of worms and at various stages of their development. The bulk are given as an oral drench but some are injectable, in bolus form or incorporated in feed blocks. Oral drenches hit the worms with one large dose while feed blocks and boluses work by exposing the ewe to low levels of wormer over a long period. Boluses are swallowed into the rumen where they release the chemical over several weeks. They are possibly best where sheep graze the same land every year. Worming through feed blocks is convenient but ewes must be used to blocks and there is a risk that some will not eat their ration.

Collect literature about wormers direct from companies (they advertise in the farming press) or from local agricultural merchants or veterinary surgeons and take time to decide on a rotation, system and product. Walking into a shop and deciding at the counter is unwise.

When choosing a wormer make sure that:

- It is clear which chemical group it belongs to.

- It is effective against the target parasite.
- It is suitable for the class of stock, such as lambs or pregnant ewes.
- The expiry date is adequate.

Work out how much is needed so as to avoid waste at the end of the year and also check the withdrawal period. Most wormers have withdrawal periods of around fourteen to twenty-one days which means that dosed stock must not be sold for slaughter until after that period. Bear in mind the withdrawal times when dosing lambs that are destined for slaughter.

Worm Control Programme

When drawing up a worm control programme temper it with the following, even though nature will do its best to spoil it:

- Plan grazing strategy at least two years ahead.
- Draw up a basic dosing programme to take into account clean grazing, weather, time of lambing and selling.
- Bear in mind other regular activities such as vaccination, shearing, lambing and weaning to rationalize handling and minimize stress.
- Always dose and move. Dosed sheep left on the same area are reinfected immediately.
- Keep dosed sheep penned for at least eight hours before turning to pasture to void any surviving eggs.
- Worms are a flock problem so the whole flock or group should be treated.
- As a principle it is better to dose early in the season to prevent a build-up of worms rather than dosing heavily at peak times in the summer.
- Worm as routine in a drought year because it reduces next year's infestation.
- Outwintered ewes and lambs can develop parasitic problems in mild winters when temperatures are high enough for eggs to hatch.

- After a dry spell rain can trigger hatching and produce mass infection but lambs may have no immunity because of the low infection levels during the drought.
- Treat tups at the same time as ewes and more regularly if they are confined to a ram paddock.
- Sheep that are wintered away should be dosed when they go and when they return.
- Monitor disease by selling lambs and cull ewes direct to an abattoir and asking if there were signs of parasitic damage.

Selection

Sheep develop some immunity to worms as they get older and some may have an innate resistance. It has been suggested that 20 per cent of ewes in a flock carry the bulk of the worms, which implies that immunity could be bred through selection. Sheep with resilience to roundworms prevent the establishment of the ingested larvae and shed less eggs.

Organic Production

Parasitic worms are a problem in the organic flock that is restricted in its use of medicines. Natural control could be by early lambing to avoid spring hatching, clean grazing or very extensive grazing.

Liver Fluke

Liver fluke disease is caused by the flatworm *Fasciola hepatica* which invades the liver causing ill thrift or death. Part of its life cycle involves a mud snail which lives in wet pasture, therefore the problem is associated with wet, reedy areas, slow-flowing streams and ditches and even wet areas around troughs and tractor ruts. Highest risk periods are after wet summers when the soil was consistently saturated – the more wet months in succession the more severe the outbreaks.

There are specific flukicides to control the disease that is caused by immature and adult fluke, but not all flukicides are effective against both stages; immature fluke cause acute disease in the autumn and adult fluke cause chronic disease in the winter. Take veterinary advice on flukicides.

Keep liver fluke under control by:

- Fencing off wet areas, streams and ditches.
- Draining wet areas.
- Strategic dosing with the correct flukicide – basically three annual doses, one each in spring, autumn and winter. On high risk farms and in wet years, give an extra dose six weeks after the basic dose.
- On small areas spray molluscicides to kill the host snail.

Tapeworms

Tapeworms causing gid and hydatid disease in sheep and humans have the dog as an intermediate host. Dogs pass the infection to humans through licking. Cysts form in any organ including the brain but usually the liver.

Control is to:

- Worm dogs regularly with a drug effective against cestodes.
- Never let dogs scavenge dead sheep.
- Avoid grazing pasture used for sheep dog trials or by hounds.

EXTERNAL PARASITES

External parasites such as lice, ticks and mites cause irritation, damage skins and fleeces and can spread disease. Most are preventable. Flystrike on soiled, damp or greasy wool is common in summer and autumn in humid weather and where trees and hedges harbour flies. The Greenbottle (*Lucilia sericata*) lays eggs on the wool and dung, and the maggots (Fig 66) that can hatch within three days, feed off the skin. Their toxins, plus physical damage, will kill within a few days.

Struck sheep characteristically wag their tail, twist around like a banana, nibble and stamp a foot (where maggots have invaded a hoof). However, in the early stages the larvae may be hard to find but if the behaviour and conditions suggest flystrike, keep looking. In the advanced stages sheep will isolate themselves, seek shade and there will be dark, moist patches on the fleece.

Clip the affected area and remove maggots, taking care that they do not escape to clean fleece. Treat the damage with insecticidal cream or oil to aid healing. Do not treat with neat disinfectant or antiseptic because these can be toxic in concentrated form. The damaged skin will need protection from sunburn.

Fig 66 The result of flystrike. Maggots invade the wool and flesh.

Fig 67 Pour-ons are sprayed along the backs of sheep to control external parasites such as blowfly strike. This system has replaced dipping on many farms. The mode of action of pour-ons varies between products, so they need to be chosen with care. They also need to be handled with care and their instructions for use strictly adhered to.

Control the problem by good husbandry such as, crutching, preventing scouring and footrot, and treating any wounds that may result from tailing, tagging and shearing. Dipping in an appropriate chemical or using a pour-on formulation (Fig 67) should prevent the problem. Pour-ons that control flystrike have different modes of control – some repel the flies and some stop the larvae developing – so they need to be chosen appropriately.

Never totally rely on preventive treatment – always watch for signs of trouble.

Sheep Scab

Sheep scab is caused by a mite that is specific to sheep and almost invisible to the naked eye. The condition is highly contagious and the mite is capable of enormous multiplication, especially in the winter. Signs are extreme itchiness, rubbing and wool loss, and prevention and cure are by dipping in an appropriate chemical or by giving an injection – which will also control lice and ticks. Prevent infection by avoiding contact with other flocks and, because the mite can survive away from the sheep for two weeks, beware of infection from shearing equipment, transport and market pens.

Scrapie

Where a ewe is rubbing excessively and is nervous and nibbling for no clear reason, suspect scrapie – a degenerative brain disease related to Bovine Spongiform Encephalopathy (BSE) and so called because of infected sheep's compulsive scraping and rubbing. Affected sheep die and currently only a post-mortem can confirm the disease. The progeny of affected ewes should not be kept as replacements.

Dipping

Dipping is an effective way to control external parasites but to some extent has been replaced by injections and pour-ons. Dip chemicals are extremely toxic and handling them and disposing of them after dipping are major concerns. In the UK the Environment Agency will advise on the latter. General advice on dipping includes:

- Follow the instructions.
- Do not dip full, hot, wet or tired sheep.
- Wear protective clothing.
- Know the volume of dip.
- Top up when recommended (sheep strip the chemical out of the dip with their wool).
- Soak to the skin (for one minute) and push the head under twice.

GENERAL HEALTH

Even the healthiest flock will lose sheep and then the carcasses must be disposed of safely and as soon as possible. Paying the local knacker's yard to collect it is the simplest in a small flock. Digesters and incinerators are available for on-farm use but burying is acceptable provided the soil is sufficiently deep and it is away from watercourses. The Environment Agency will advise. Cover the carcass with quicklime or a powerful disinfectant before replacing the soil.

Shepherd's Health

Healthy sheep need healthy shepherds. Diseases which are transmissible from sheep to humans (zoonoses) include infectious abortions, orf (a highly infectious viral disease resulting in skin pustules), salmonella, hydatids and anthrax. Doctors should always be told when a patient has been in contact with sheep.

Accidents to shepherds arise from treading, butting, lifting, climbing and injecting. Chemicals such as sheep dips, sprays and drenches must always be handled with care and, when advised, used with protective clothing.

Environmental hazards include skin cancer, infected cuts and tetanus. Shepherds should always have an up-to-date tetanus inoculation; most are conscientious about vaccinating their sheep against tetanus but forget to protect themselves.

Health Records

Keeping records of health problems and treatments will identify areas where health control programmes are needed. Brief details should also be included on the individual sheep records to identify any weakness or to trace family susceptibility.

Facts to keep on a flock health record include:

- Date.
- Identity of animal.
- Symptoms.
- Diagnosis.
- Treatment.
- Outcome.
- Comments.

In the UK an Animal Medicine Book must be kept to record the purchase date, name, quantity and batch number of medicines plus the identity of the animals, the start and finish date of treatments and the withdrawal period.

9 Lambing ────────────────────

During the six weeks before lambing there are a number of problems to look out for:

- Abortion.
- Metabolic disease.
- Prolapse.

Most flocks suffer abortion caused by infection, rough handling or barging at the feed trough. A normal incidence is 1–2 per cent.

Initially any abortion should be treated as infectious, which means isolating the ewe, handling her and the aborted material hygienically and seeking veterinary diagnosis. The veterinary surgeon will need the fresh foetus and foetal membranes. He will also need to know:

- The expected lambing date.
- The feeding regime.
- Any history of abortion.
- Affected age group.
- If any ewes have been bought-in.

Enzootic abortion and toxoplasmosis account for 80 per cent of abortions and both are controlled by management and vaccines. The veterinary surgeon will give advice on both as will any good veterinary book (*see* Further Reading).

Suspect infectious abortion where there are:

- Stillborn premature lambs.
- Abnormally dark-coloured stillborn lambs.
- A high barren rate.

- A high proportion of small and weakly lambs.
- A number of weakly lambs born several days early.

Pregnant women should never be involved with the lambing flock because they risk infection. Nor should aborted ewes be used as foster-mothers until infection has been ruled out.

Metabolic Diseases

In the last four weeks of pregnancy the foetus makes 80 per cent of its growth and metabolic diseases occur when the ewe has to provide for this growth faster than she can metabolize nutrients from her body. The problems are commonly

- Pregnancy toxaemia (twin lamb disease).
- Hypocalcaemia (lambing sickness).
- Hypomagnesaemia (staggers).

A ewe lagging behind the flock, staggering, reluctant to feed or showing signs of nervousness is suspect.

Pregnancy toxaemia usually occurs in the last few weeks before lambing. Hypocalcaemia strikes close to lambing or in early lactation and can be triggered by the stress of moving or transporting. Hypomagnesaemia is caused by magnesium imbalance and usually occurs in early lactation on fresh, heavily fertilized grass.

All are preventable by correct feeding and avoiding stress, and can be treatable if recognized in the early stages.

Checklist of Items Commonly Needed for the Lambing Season

General

Lambing record sheet

Phone number of veterinary surgeon

Lamb / afterbirth disposal system – Beware carcass disposal regulations

Disinfectant – For cleaning mothering pens, etc.

Strong plastic bags / sack – For collecting afterbirth, dead lambs and foetal material

Thermos flask / supply hot water – For making hot drinks or (cooled) for thinning thick colostrum for tubing or mixing dried colostrum

Shelter area for ewes and lambs – Cheap housing, hurdles, windbreak material, bales.

Shelter area for shepherd – Caravan, hut

Bedded area ready for pet lambs – Pen, deep box

Notebook / pencils that can be used in the rain – For lambing data, reminders and lists

Fox deterrents – Electric fence, bells, flashing lights, etc.

Torch – Remember batteries if not rechargeable

Spare bulb / batteries – For torches

Large dustbin with lid – Many uses including holding concentrates or for keeping foster and natural lamb during fostering

Very sharp knife – For skinning lambs when fostering

Baler twine and pocket knife – Invaluable

Plastic bags – For rubbish

Buckets – Can never have too many

For the Ewe

Food, water and hay dispensers – If penned after lambing

Prolapse retainer / harness

Medicines for pregnancy toxaemia, hypocalcaemia and hypomagnesaemia

Hoof and dagging shears, worm drench and gun, footrot spray – For routine treatment if practised

For Assisted Lambing

Long-acting antibiotic – For the ewe

Water, soap, towel and mild disinfectant – For clean hands and arms

Lambing lubricant – Pure soap flakes or proprietary type

Lambing ropes – Have several so that there is always one sterilized

Antiseptic and plastic containers with lids – To store sterilized equipment

Source of boiling water – For cleaning and sterilizing

Syringes – 5ml, 20ml, 50ml

Needles – 16 gauge × 12.5mm and 19 gauge × 25mm

Arm length poly gloves – For hygiene

Rubber gloves – For grip

For the Lamb

Iodine or antibiotic spray – To treat navels

Towels / absorbent material – For drying lambs if necessary

Thermometer – To check for hypothermia. If digital check the batteries

Scissors – To trim umbilical cord

Warming box and heat source

Colostrum – Frozen from ewe or cow or dried

Colostrum container – To milk and carry colostrum

Stomach tube and 50ml syringe – For tubing colostrum

Glucose solution – For intraperitoneal injection

Lamb jackets – In case of turnout into bad weather

Cardboard boxes – For transporting lambs, temporary home for a foster lamb, makeshift intensive care and 1000 other uses

Ewe milk replacer plus mixing utensils

Feeding bottle with teats – For topping up

Bucket feeder, teats, tubes – For artificial rearing

Thermos to carry warm milk for topping up, etc.

Glucose and electrolyte – For scouring lambs

Antiseptic / antibiotic spray – For wounds or injection site

Liquid paraffin or soapy water – Enema

Rubber or plastic gloves – For handling material which might be infectious

Castration and tailing kit – If practised

Lamb tags – For identification

Marker spray / paste – To identify lambs and dams

Fig 68 Hurdles make good mothering-on pens. Standard hurdles with bars need the lower bars close together to prevent small lambs escaping. Solid sides – such as plywood – are the most secure and prevent ewes in adjoining pens from being distracted by a neighbouring ewe, lambs and food. Sacking or plastic fertilizer bags tied onto hurdles make them more secure but any loose material or string is a magnet to baby lambs. Pens can be made from small straw bales but must be well secured.

Prolapse

Prolapse of the vagina may occur around three weeks before lambing usually in older, fat ewes. The first sign is the pink fleshy

Fig 69 Old plastic feed or fertilizer bags make cheap hay feeders for individual mothering-on pens. Cut slots in the bottom, stuff with hay, tie the neck tightly and suspend it well clear of the ground.

vagina showing through the vulva when a ewe is lying down. It may disappear when she stands and may never get any worse. If it stays visible it may need a truss, plastic retainer (Fig 74) or stitching to hold it in place – be guided initially by the local veterinary surgeon. Treated ewes need watching so that stitches or retainers can be removed at lambing, although some are able to lamb with them in place. Since the problem is likely to recur, prolapse cases should be culled and, because the weakness is believed to be heritable, their progeny should not be kept as replacements.

PREPARATION FOR LAMBING

Good preparation is vital for a successful lambing. In the last six weeks jobs may include housing, clostridial vaccination, worming and crutching. Ewes should be grouped according to their lambing dates, if practical, so that feeding is targeted accurately. Very thin or very fat ewes may have separate groups for individual feeding.

In the last two weeks the lambing facilities and equipment (*see* box) should be in place. Mothering pens (Fig 68) should be under cover or in a sheltered area. Allow one for every six to ten ewes in the flock or

Fig 70 Lambing cubicles are used in America. The cubicles are used with confined and housed flocks where ewes have little space and no privacy at lambing. They are sited on the perimeter of a pen, away from working areas, and well bedded. Ewes looking for a birth site go in voluntarily and lamb in peace. The low threshold on one side allows the ewe in, but prevents new-born lambs getting out. They will need to be well cleaned between occupations and ewes will need to be watched during lambing because they risk lying on their lambs.

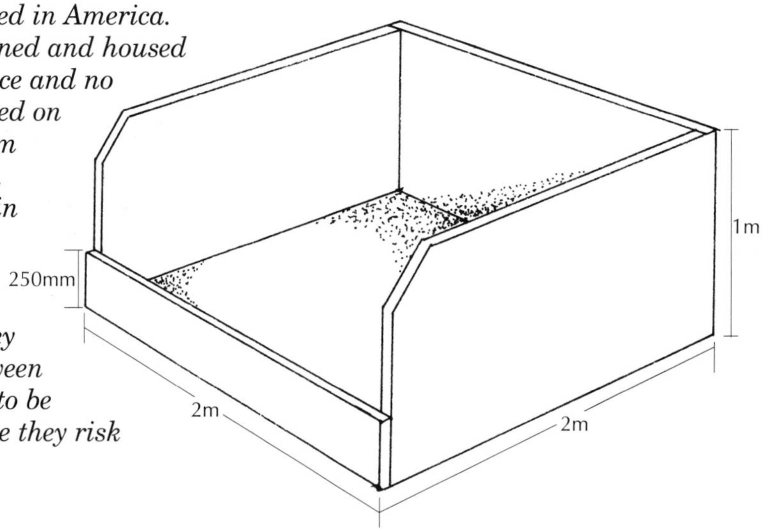

one per three ewes in a synchronized flock, plus a pen for hand rearing surplus lambs. Pens must be secure, have easy access, be at least 1.5m × 1.5m and 1m high with no gaps where lambs can be trapped or escape, no projections and no loops of twine to ensnare lambs. Prolific flocks need larger pens to accommodate large litters. Old plastic feed bags (Fig 69) make effective individual hay feeders.

Shepherds in America use lambing cubicles (Fig 70) in housed flocks. Sited away from a busy area, the ewes go in voluntarily to lamb in privacy. The low threshold on one side lets the ewe in, but retains the lambs.

Dig a disposal pit away from a water course for afterbirths and dead lambs and cover it with a heavy top that is considerably larger than the pit to prevent scavenging.

Outside Lambing

Ewes lambing outside should move to the lambing area at least ten days before lambing to allow them to settle. The area should have been sheep free for a few weeks to let it freshen up. Provide shelter from all wind directions and check dangers such as fencing and gates that new-born lambs can get through but ewes cannot follow.

The Shepherd

Have light, warm clothes with loose sleeves for rolling up. Waterproofs are necessary for outdoor lambing and waterproof leggings indoors protect trousers from kneeling and handling sheep but some waterproofs are noisy to walk in and can spook sheep. File nails, remove rings and have somewhere comfortable to sit.

- Adopt a routine for bedding, watering and feeding. Look at the flock regularly – every two to four hours by day depending on the amount of activity. Lambing may last from three to six weeks so plan a sensible night routine such as taking a last look at midnight and a first look at 6am. In large flocks a twenty-four-hour watch (three eight-hour shifts) pays off in terms of lambs saved and reared.

In addition:

- Have easily available snacks, meals and hot drinks.
- Have easy access to a sink and hot water.
- Have all lambing equipment close at hand.

- Tell the veterinary surgeon when lambing starts.
- Have a list of expected lambing dates for individuals and, if scanned, the number of lambs expected.
- Have record sheets (Table 7) to keep up to date during lambing and a notebook and pencil to avoid relying on memory.

Lambing Times

Lambing is not spread evenly over twenty-four hours. In a large flock it will peak at 6–7am and 6–9pm with a lull from midnight to 3am. However, work in America and the UK suggests that flocks can be persuaded to lamb during the day; the theory being that a disturbance, such as feeding, inhibits labour. Flocks which were fed at around 9–10pm resulted in 90 per cent of lambs being born during daylight hours. The key is to feed the ewes at the same time each day, starting the routine three weeks before lambing to condition the flock.

LAMBING

The ewe goes through several stages when she is lambing:

- Some twenty-four to forty-eight hours before lambing the vulva is very pink

and slackens as the muscles relax. The udder tightens and the teats distend.
- Within twenty-four hours she may become restless, distracted and occasionally paw the ground. She may steal a new born lamb from another ewe and cause confusion.
- Within four hours of lambing she will try to isolate herself, paw the ground and find a birth site – usually on the perimeter of the lambing area.
- Serious attempts to lamb begin when she lies on her side, characteristically lifting her head and pointing her nose skywards (Fig 71). The cervical seal appears from the vulva as a creamy mucus and she gets anxious, bleats and gets up and down licking the ground where amniotic fluid has been spilled. Uterine contractions move the lamb towards the cervix, leaving a hollow forward of the pelvis. This stage may last three to four hours.

Lambing Sequence

Lambing proper begins when she is visibly straining and membranes and mucus appear from the vulva. The major membrane is the water bag containing the foetal fluids (Figs 72 and 73).

She strains more urgently, the waterbag bursts to lubricate the birth canal and the lamb's nose and front feet should become

Fig 71 The start of lambing. A ewe lifts her nose towards the sky.

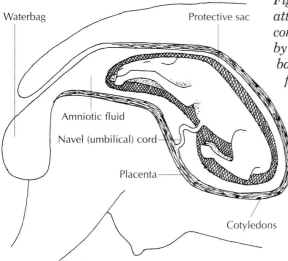

Waterbag

Protective sac

Amniotic fluid

Navel (umbilical) cord

Placenta

Cotyledons

Fig 72 Early stages of birth. The lamb is attached to the placenta through the navel cord, and the placenta is attached to the uterus by button-like cotyledons. There are two water bags – the first one containing the amniotic fluid and the second a protective sac around the foetus. As the lamb is expelled the first water bag lubricates the way for the lamb. Sometimes the second water bag remains around the lamb until it shakes its head free or the ewe licks it. The navel cord will break during birth or when the ewe stands up.

Fig 73 (Below) The waterbag is the first membrane to appear before the lamb is born.

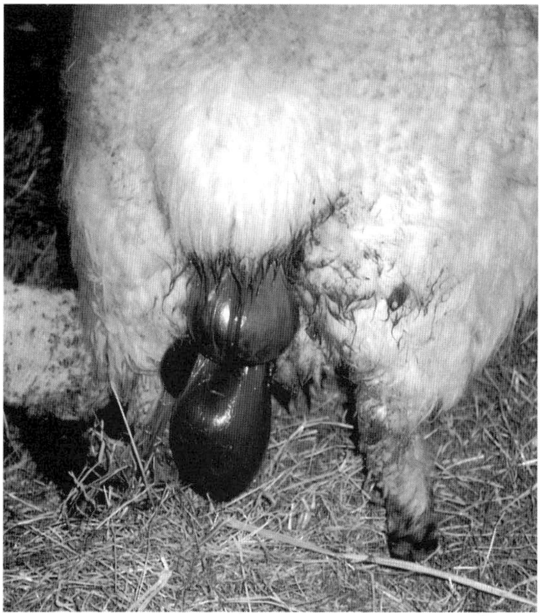

visible. After a few minutes of serious straining (a lot longer if she is restless) she should expel the lamb.

The umbilical cord breaks naturally and the lamb shakes its head free of membranes and coughs and breathes.

The ewe licks it around the face to clear membranes and dry it; when the lamb tries to stand she nudges it towards a teat.

After nursing the first lamb she may become distracted, paw the ground and lie down to lamb another.

Time Scale

Predicting the time for the birth process is an inaccurate science because all sheep differ but it is useful to know what to expect.

From the:

- Start of serious straining to the appearance of a water bag 30 minutes
- Water bag to appearance of lamb 20–30 minutes
- Birth to the lamb getting on its feet 10–20 minutes
- Birth to suckling 30 minutes

- From birth of first lamb to birth of second lamb 30 minutes
- Birth of first lamb to the expulsion of afterbirth 2–3 hours (half an hour to eight hours is normal)

Ewes lambing for the first time are likely to take longer than experienced ewes.

118

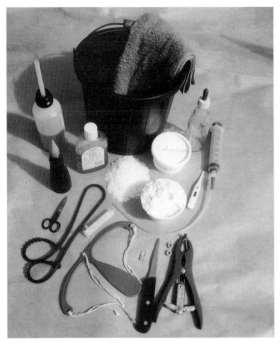

Fig 74 Basic lambing kit. Clockwise from bottom right: tailing / castrating pliers plus rings; sharp knife for skinning dead lamb for fostering; lamb wing tag; prolapse retainer; syringe; lambing rope; scissors for trimming navels; iodine; lambing lubricant; disinfectant; bucket and towel; soap flakes to help lubricate hands; milk powder; frozen colostrum; feeding bottle and teat; stomach tube; digital thermometer.

During the Birth

When a ewe starts to lamb leave her undisturbed and check her records to see if she has a history of problems. Make sure that the basic lambing kit (Fig 74) is handy:

• Lambing gel or solution of soap flakes as lubricants.
• Warm water and clean towel.
• Mild disinfectant.
• Lambing rope.
• Scissors.
• Have a mothering pen ready with water and clean bedding.

• Keep an eye on her. Lambs can be born in tough membranes which prevent breathing or they can come backwards and suffocate in the cervix. Both cases need quick action to free them.

Assisting a Birth

Some ewes need help at lambing – usually fat ewes with large lambs. Help will range from a gentle pull when a lamb is large, to full blown assistance where a lamb is wrongly presented. A reasonable goal is 10 per cent of ewes assisted. This means that on average 90 per cent of ewes do not need assistance, so it is not a major issue. Colleges or training organizations run lambing courses using simulators, and when a ewe with a lambing problem is taken to the veterinary surgeon he should be asked to explain what he is doing.

The skill is:

• To know when to investigate.
• To know when to assist.
• To know how to assist.
• To know your limitations.
• To seek veterinary help sooner rather than later.

If a lamb spends too long in the birth process it can separate from the placenta and die before birth. Therefore protracted births are risky.

The consensus is that investigation should be left until:

• Around 90 minutes after the first signs of straining and no lamb is visible.
• Or 30 minutes after the water bag has appeared and no lamb is visible.
• Or a ewe appears to give up vigorous straining, loses interest or seems exhausted.

Have a system for quietly cornering and catching a ewe; they can be lured by a borrowed baby lamb or even by a towel lying on the ground. A quiet ewe can be approached while she is preoccupied with straining. Ewes can be restrained and investigated on their side or standing up.

Initially make a cursory check to see if the presentation is correct. If, for example, both legs are back (Fig 78) it is easier to correct it before the lamb is hung (Fig 79). With a scrubbed, disinfected and well-lubricated hand, cup the fingers and thumb to form a cone, slide it into the vagina a few inches and check the cervix. If the cervix does not seem to have dilated and there is no entry into the uterus, massage it gently for a minute and give it more time. If it refuses to open, and the ewe is a maiden, it may be a condition known as ringwomb and will need veterinary attention.

If the cervix is open and a lamb can be touched it should be possible to identify instantly if it is normal presentation (Fig 75) by the presence of a nose and two hoofs. If it is normal but she makes no progress after another fifteen minutes help to draw (pull) it because it is likely to be large or she is tired.

Drawing a Lamb

The basic rules for drawing a lamb are:

- Disinfect and lubricate hands and arms.
- Ease a handful of lubricant into the birth canal and around the head of the lamb.
- Pull on as many parts of the lamb as possible – that is, both legs and the head. Straighten the lamb's front legs and draw them alternately. Always be gentle but firm. Only experience can teach how hard to pull.
- Normally draw slightly downwards towards the udder.
- Pull when the ewe is pushing (straining).
- To help free a tight lamb rotate it slightly, or roll the ewe onto her other side.
- Rubber gloves give a comfortable grip.
- If the feet are secured and pulled with a lambing rope make sure that the nooses

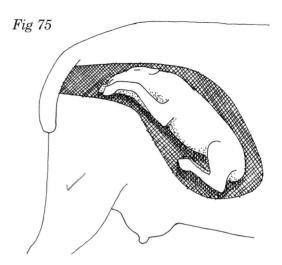

Fig 75

Normal presentation of a single lamb.
If a ewe is straining unsuccessfully to deliver a normal presentation it is likely to be too large for her. Straighten the lamb's legs and draw the lamb while easing the fingers around the back of the head – using lots of lubricant – to push it out. Pulling the legs alternately and rotating the lamb gently one way and then the other may help to release it. Wearing a rubber glove helps to grip legs when they need a strong pull.

are above the fetlock joints and not around the pastern joints.

If there is no sign of a lamb, or if part of a lamb is visible but is making no progress, gently, hygienically and with lots of lubricant investigate further into the uterus. Be gentle because the uterus lining is fragile, and feel along the backs of the lambs to avoid disconnecting the umbilical cord.

If the contents of the womb feel naturally lubricated, and there are no brown unpleasant discharges, there is probably no urgency. Decide what the problem is (Figs 75–83) and correct it or seek experienced help.

Lambing is a clean activity so smelly mucus or membranes may herald a dead lamb or an abortion. A dry foetus also suggests a dead lamb. Withdraw it with care,

Fig 76

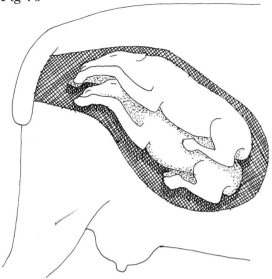

Normal presentation of twins. *These would normally be lambed in an orderly sequence. If they come together and get stuck, ease one back and draw the other. They are less likely than a single to be too large.*

Fig 77

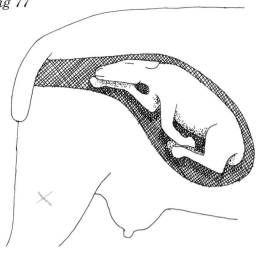

One leg back. *This is a common problem. In the case of a small lamb or a 'roomy' ewe (not a maiden) it is possible to draw it with one leg back. Otherwise correct the situation as suggested in the presentation of both legs back (Fig 78).*

Fig 78

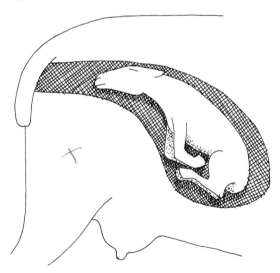

Both legs back. *In this case the lamb loses its natural arrow-like shape and sticks at the shoulders. Ease the lamb back into the uterus, cup the feet in the hands and draw them forward individually, and the lamb should be expelled in the natural lambing position. Early recognition of this problem will avoid a hung lamb (Fig 79).*

Fig 79

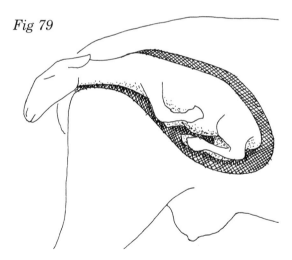

Hung lamb. *This happens when both legs are back and the ewe has managed to expel the head. Often the head is swollen and the tongue blue but this does not mean that the lamb is dead. If the head is dirty, wash it before returning it slowly to the uterus which is best done by putting the ewe on her back, lifting her rear end up and supporting it against a bale. It will take copious amounts of lubricant and patience. Correct and lamb as for both legs back (Fig 78).*

Fig 80

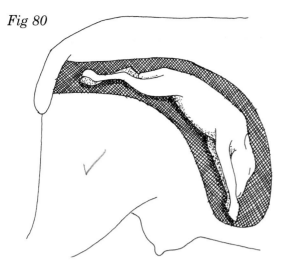

Fig 81

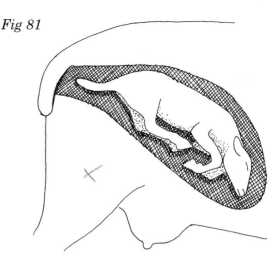

Backwards. *This is common with multiples, when a ewe may lamb it unaided. But if it gets halfway out and the ewe stops for a rest it will suffocate. As soon as the whole tail is visible it must be drawn quickly because the navel cord will be stretched and trapped against the pelvis. Pulling one leg at a time can narrow the lamb's pelvis, and rotating through 45 degrees will help. If the lamb is also upside down draw it straight out and not towards the udder. Swing the lamb, and clear the mucus from the nose and mouth. The danger with a lamb being born backwards – especially a large one – is that the ribs can break.*

Breech. *Usually identified quickly because the tail is visible and often the ewe stands in a characteristically rumped position. Correct it to the backwards position by holding the hock, pushing it gently upwards and forwards, then cup the foot and ease it into the birth canal. Do the same with the other leg and deliver it in the backwards position.*

Fig 83

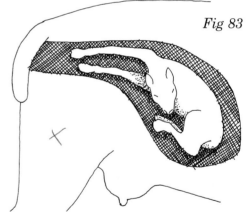

Fig 82

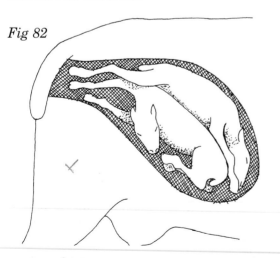

Head back. *This is one of the most difficult to correct. Put ropes on the legs and return them to the uterus. Identify the head, straighten it to lie between the forelegs and rope it because it inevitably springs back to its original position. Ease head and legs into the birth canal and lamb as for normal presentation.*

Wrong presentation of twins. *In this case the backward presentation is likely to be the easier to lamb and is best drawn first if the lower lamb can be eased further into the uterus. The key is to match the legs with the lambs, and identify the front and back legs (Fig 84) so as to distinguish which is the backward presentation.*

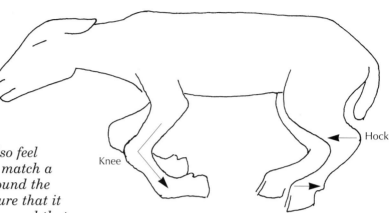

Fig 84 One problem with an assisted lambing is identifying the back and front legs. The drawing shows the essential differences; in the front leg the joints bend in the same direction; in the back leg they bend in opposite directions. The knee joint and hock also feel different. It is also vital to match a pair of legs; feel closely around the inside of one leg to make sure that it actually joins the second leg and that they both belong to the same body.

feel for other lambs and treat as an infectious abortion.

Whenever a hand has been put into the uterus the ewe should have an antibiotic injection or pessary, and the difficulty should be recorded so as to identify a recurring problem or persistent offenders.

Professional Help

If the ewe is taken to the veterinary surgeon, take with her:

- The navel iodine (*see* later).
- A towel to assist drying on the way home.
- A shallow strawed box for the lambs.
- A stomach tube to feed colostrum if the journey is long.
- A foster lamb if dead lambs are suspected.

Routine for the New Lamb

As each lamb is born there should be a routine to make sure nothing is forgotten.

- Clear mucus from nose and throat with finger and check that it is breathing by watching the chest expand and contract.

- If it is slow to breathe hold it securely by the back legs and swing it through 180° three or four times. Other methods to encourage breathing include putting a piece of straw up a nostril; rubbing vigorously with straw or a towel; pinching the skin; using an oral spray that stimulates the heart and breathing. Sometimes it may take several minutes to get a lamb to breathe regularly.
- Shorten the torn umbilical cord to around 7cm with scissors. Never cut the cord before it has torn free because it can haemorrhage.
- Treat the navel immediately with veterinary iodine solution (preferably 10 per cent). The navel is an open wound and must be protected against disease organisms. The iodine can be sprayed, shaken on from a sauce bottle or dipped in a small container. Complete immersion is best because it surrounds and soaks the cord at the point of entry.
- Check for milk by gently drawing some from the teats. The plug in the teat canal (*see* Appendix I, Fig 116) will offer some resistance. A ewe should be milked while she is standing.

Colostrum

The new-born lamb is totally at the mercy of its environment. It has two enemies:

- Cold and infection.

The one defence against both is colostrum (first milk) that is high in both energy and antibodies.

At birth a lamb has to survive the transition from a temperature in the uterus of 40°C to the outside temperature that could be freezing. To do this it has a layer of brown adipose fat along the back and neck which metabolizes rapidly after birth to produce energy. However, it only provides enough heat to keep the lamb warm for about half an hour and after that it must get its warmth and energy from colostrum. This is the first reason why lambs must have colostrum quickly after birth.

Unlike other animals, antibodies from the ewe's bloodstream do not cross the placenta and a lamb is born exposed to disease. Bacteria enter through the navel cord and the mouth and the only defence are the antibodies in colostrum. Initially the stomach has a high pH so as not to destroy these antibodies, but equally it does not destroy bacteria.

The lamb's intestines are lined with cells that allow antibodies from the colostrum to pass straight into the bloodstream. These cells are slowly replaced after twelve hours and may be gone by twenty-four hours. Although the ewe continues to secrete antibodies for around four days after lambing, the lamb loses the ability to absorb them, which is the second reason why lambs must have colostrum quickly after birth.

Colostrum is also a laxative and is essential in getting the gut working to prevent watery mouth. So the rules are:

- Lambs should have some colostrum within half an hour of birth.
- Do not give ewe milk replacer before colostrum because it reduces the time that the intestine can absorb antibodies.
- A lamb needs 200–250ml of colostrum per kg of bodyweight during the first fifteen hours. So a 5kg lamb will need more than one litre.
- Most healthy ewes will have enough colostrum for their lambs.
- If in doubt – for example a last-born triplet – feed extra by stomach tube.
- Feed 50ml per kg of bodyweight for a first feed. Do not overfeed if it is reasonably vigorous because it can make it drowsy and fail to suckle while it is still able to absorb antibodies.
- Have enough spare colostrum for 5 per cent of expected lambs at 200ml per lamb.
- Never be without spare.

Sources are:

- Colostrum milked from an older ewe with a single lamb.
- Goat colostrum (must be free from Caprine Arthritis Encephalitis).
- Cow colostrum. There will be no appropriate antibodies (unless the cow has a clostridial vaccine before calving) and a slight risk of inducing anaemia. It is best to have a mixture from several cows.
- Commercially available dried colostrum.

Colostrum can be frozen in ice cubes or small plastic containers and kept for a year. Defrost gently in a container in warm water. Boiling or microwaving destroys antibodies. Feed at body temperature of 39–40°C.

- Watch to ensure that the lamb sucks – wagging tails and sucking noises are a guide but they may just be pulling on a staple of wool. Check the teat after suckling to see if it is soft and damp. It is normal for some ewes to be restless at this stage and keep circling the lamb, making it difficult for it to latch on. If a lamb is slow to find a teat, guide its head to it while nudging and pushing the base of its tail.

  It is normal for a ewe to paw a lamb aggressively to make it stand up. If a lamb is slow to stand it may be because the joints are weak or the legs are bent from having been in an awkward position in the uterus. The lamb may need to be held to suck but usually strengthens within a few hours.

- Let the lamb have its fill, but if it is small, and siblings are expected, milk off some of the colostrum (see panel) into a clean container and share it among them later by stomach tube (Fig 85). Any weakly or small lambs should have colostrum by stomach tube within twenty minutes of birth.

- Make sure meconium (first content of the bowels) is expelled. It is very dark and sticky. If none is seen and after a few feeds the lamb seems to be limp or slobbering give an enema of glycerine, liquid paraffin, or up to 20cc of soapy water flushed slowly into the rectum at body temperature from a small syringe (without the needle) or the stomach tube. It should expel meconium almost immediately but if there is no improvement suspect watery mouth (see later).

- Check the lambs for in-turned lower eyelids (entropion), indicated by a weeping eye. Pull the skin below the eye to flip it out; correction at this stage may avoid surgery later. Note any cases on the ewe records because it is possibly inherited.

- Expect more lambs until the afterbirth (placenta) appears (Fig 86) then watch for it to be expelled. If no afterbirth is seen (she may have eaten it or it may have been scavenged) suspect it has been retained. Normally this is not harmful but keep an eye on her health and be prepared to give an antibiotic.

  Never pull afterbirth away, but if it is slow to cleanse it can be shortened by scissors to reduce the chance of infection. Once cleansed, dispose of it hygienically.

- Leave the family on the birth site for as long as feasible to improve bonding. Do

Fig 85 To use a stomach tube sit the lamb comfortably on your knee, coat the tube with vegetable oil, let the lamb chew your finger then slide the tube over the tongue and into the stomach until about 5–8cm are left. If the tube goes into the lungs by mistake the lamb will cough violently. Normally it will relax and chew the tube. Attach the full syringe to the tube and empty gently. When finished, pinch the end of the tube and withdraw it slowly.

125

Fig 86 The afterbirth signifies that she has finished lambing. Characteristically it contains red buttons or cotyledons on the membrane.

will accept any which are taken away and returned during that period.

- Keep ewe and lambs in a mothering pen for 24–48 hours. Make sure that the lambs are feeding (especially any small, weaker ones), and free any tails which get stuck down over the anus with faeces.
- Record the details of the lambing with particular emphasis on any lamb losses so that the reasons and any trends can be identified and corrected (*see* box on losses).

Never be surprised if siblings are different breeds. Ewes are capable of being fertilized by several rams and bear littermates with different sires. Lambed ewes can also have a second lamb several days or weeks after the first; this is a well-documented but unexplained phenomenon.

After twelve hours the ewe could be:

- Worm drenched.
- Foot trimmed.

After twenty-four hours the lambs may be:

- Tagged.
- Weighed.
- Tailed.
- Castrated.

not distract the ewe by feeding her until she has cleaned and fed the lambs. Offer warm water immediately; ewes lose at least a gallon of body fluids at lambing and their thirst is immense.

- Put the family into a mothering pen. Most ewes follow their lambs readily if carried slowly and low to the ground. If contact is lost the ewe will run back to the birth site.
- If a lamb is taken away for treatment the ewe may reject it when it comes back, so keep the afterbirth to wipe on the lamb when it returns. It may take six to ten hours for a ewe and multiple lambs to bond, so there is a better chance that she

Families should be identified before they leave the pen so that they can be matched up in the field. Write the same number or mark on one side of each member of the family with a scourable spray or paint. A spot on the back of the neck identifies a lamb as a single.

Pedigree breeders will also identify the lambs with ear tags or tattoos as required by their breed society. For the commercial producer the poultry wing tags (Fig 88) are easy to use and usually stay in place long enough to identify lambs at weaning when they can be given permanent tags if necessary.

When a family is turned out from the pen it should be added to a small group to minimize the chance of mismothering.

Lamb Losses

In every flock there will be losses and lambing is the riskiest time. Average ewe losses are around 2 per cent.

Neonatal lamb losses of around 5 per cent is a good target in a small flock; 10 per cent may be acceptable, but any higher suggests a problem that needs identifying. Hence a record of losses and reasons should be kept for analysis (*see* box p.135).

About three-quarters of lamb deaths occur at birth and in the first four days, so this is the time to concentrate effort on caring for the lambs. More lambs die from management failures than infectious disease or dystokia.

Prevention

- Correct feeding of the ewe to produce viable lambs and good colostrum.
- Supervision at lambing.
- Adequate colostrum.
- Treat the navel.
- Hygiene – especially bedding.
- Shelter.

Birth Weight

Neonatal mortality increases as birthweight falls below about 3kg putting lambs at risk of hypothermia and starvation. At the other end of the scale large lambs are at risk from dystokia. Figures suggest that in modern breeds lambs below 2kg birthweight have a 90 per cent mortality rate while lambs at the optimum weight of 4–5kg have 10 per cent mortality.

Typical birth weights are:

- Single   5kg
- Twins   4kg
- Triplets   3kg

Where there are a number of small, weak lambs at birth get veterinary advice. They may be a result of infectious abortion, inadequate nutrition or a trace element problem. Vitamin E is thought to be particularly important to young lamb survival and some compounders are adding it to their ewe feeds.

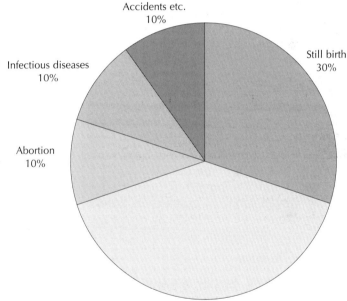

Losses of Young Lambs

Accidents etc. 10%

Still birth 30%

Infectious diseases 10%

Abortion 10%

Starvation and exposure hypothermia 40%

Fig 87 Statistics show that most neonatal lamb deaths are due to starvation, exposure and hypothermia – all preventable by good management.

Fig 88 Poultry wing tags are numbered and are used to identify lambs. Put them in like safety pins, avoiding blood vessels and leaving enough space for the ear to grow. Have clean hands, dip the tag in a mild antiseptic, push it through the ear decisively, and the lamb does not even flinch. A few lambs manage to lose them, so one in each ear is a safeguard.

Clean and disinfect the pen or if this is not possible bed deeply with more straw. Bedding can be sprinkled with granules of a chemical which vaporize and disinfect the material. Lambs are susceptible to infection via the navel (joint-ill) and the mouth (watery mouth) and the risks increase as lambing progresses and bacteria build up.

Health Checks

Problems to look out for in young lambs include:

- Hypothermia (*see* box and Fig 89).
- Watery mouth.
- Scouring (diarrhoea).
- Joint-ill and navel-ill.

In watery mouth (rattle belly) the mouth is cold and drools clear saliva, the stomach is distended by gasses and the lamb listless. It is caused by *E. coli* bacteria, ingested from wool or bedding, that stop gut movement and then cross the wall of the gut and infect the whole lamb. It is treated with antibiotics, an electrolyte given by tube, and an enema to get the gut working. Do not feed colostrum but try 20–40ml of natural yoghurt by stomach tube. Clean bedding and early colostrum should prevent the condition but in the event of an outbreak it can be controlled by oral antibiotics.

Yellow scours is likely to be caused by excess milk but grey or green watery faeces with a tinge of blood need veterinary advice. An electrolyte given by tube will help to rehydrate a scouring lamb.

A swollen navel suggests navel-ill and swollen painful joints suggest joint-ill and should be referred to the veterinary surgeon.

Fig 89 The onset of hypothermia, brought on by hunger as a result of mismothering. A number painted or sprayed with a stock marker on both the lamb and the ewe makes sure that they are quickly reunited.

Hypothermia

Exposure and starvation resulting in hypothermia (lower than normal body temperature) is the biggest single killer of lambs.

The condition is preventable by providing shelter, ensuring that every lamb gets colostrum at birth and continues to get enough milk. All lambs are at risk – even those that appear to have been successfully suckled and mothered on.

Any lamb with an anal temperature of less than 39°C is suspect. Inactivity is an early sign. Always take the temperature of a suspect lamb. Mild hypothermia makes a lamb physically unable to suck and no amount of patience will persuade it; therefore, without a temperature check, there is a risk that it will be dismissed as not being hungry.

Risk periods and causes are:

- Birth to five hours due to loss of heat.
- Five hours onwards due to starvation.

Hypothermia is treated by giving warmth and food according to age and cause (*see* following table). For example, lambs over five hours old are likely to be hypothermic due to starvation and have a low level of blood glucose. Warming them before feeding them causes fits and death.

Warming a Lamb

Always dry a lamb with a towel before warming to avoid more heat loss through evaporation. Lambs wet with birth fluids are difficult to dry and can be washed in warm water first.

Follow by gentle heat such as a hair dryer, a fan heater or even a low temperature oven with the door open. Avoid direct heat and warm the air not the lamb. Cardboard boxes with a straw bed make excellent containers.

Check that lambs are not overheated and killed by hyperthermia. Air temperature around the lamb should be no more than 35–37°C.

Special warming boxes are available but a small flock should not warrant one. A substitute is a 3kw domestic fan heater under a raised mesh floor. With any warming system there is a risk of fire.

An infra-red lamp suspended 1.3m over a box is better for keeping a lamb warm, rather than warming it.

Aftercare

A consequence of treating a hypothermic lamb is that it may be rejected by its dam when it is returned to her. If possible it should be treated in the company of the ewe and bedded on straw over a hot water bottle or under an infra-red lamp. A low partition across the corner of a pen provides a safe area.

A lamb should not be returned until it is sucking vigorously, standing, and its temperature has returned to normal. Shivering is a sign of recovery – lack of shivering is a sign of cold.

Identify the cause of the hypothermia and make sure that it does not recur. A lamb is unlikely to survive it for a second time because its energy reserves will be very low. Look for poor milk supply, sore teats, an inattentive mother or a sluggish lamb.

(continued overleaf)

Hypothermia *(continued)*

Temperature	Age	Ability	Action
37–39°C	Any	Can swallow	Dry, feed colostrum by stomach tube, give shelter with ewe. Feed a proportion of their total daily requirement little and often, starting with 50–100ml.
less than 37°C	–5hr	Can swallow	Dry, warm to 39°C then feed colostrum and return to ewe and shelter
less than 37°C	+ 5hr	Head raised Can swallow	Feed colostrum by tube, dry and warm
less than 37°C	+ 5hr	Head down Cannot swallow	Give intraperitoneal injection of glucose (Fig 90) then warm. Tube colostrum when conscious.

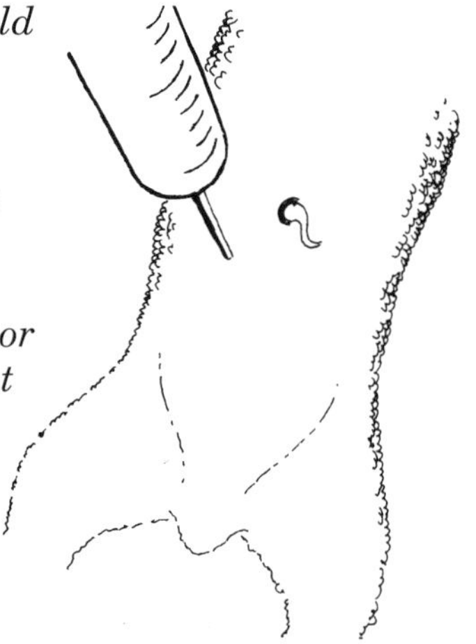

*Fig 90 **Intraperitoneal injection**. This technique should only be used to revive hypothermic lambs which are unconscious and therefore cannot be fed with a stomach tube. Inject a 20 per cent glucose solution at blood temperature at the rate of 10ml per kg bodyweight. Inject with a 50ml syringe and a 2.5cm × 19 gauge needle. Identify the injection site 10mm to one side and 25mm behind the navel. Treat the spot with an antibiotic spray or iodine, hold the lamb by the front legs, insert the needle at an angle of 45 degrees (pointing towards the rump) and empty the syringe. The solution goes direct into the body cavity and is quickly absorbed. There should be no resistance – withdraw if there is. It is not unusual for the lamb to urinate.*

Both are caused by bacteria and need antibiotic therapy. They are usually the result of poor hygiene and inadequate colostrum.

Problems to look out for in ewes include:

- Discharges from the vulva.
- Udder problems.
- Prolapse.

Discharges from the vulva after lambing are not unusual. If they are not smelly, not excessive and the ewe is not off-colour there is probably no problem. Possible serious problems are the retention of a dead lamb, or metritis (inflammation of the uterus) after an assisted lambing. These latter will need urgent veterinary attention.

Signs of udder problems are hungry lambs and ewes reluctant to suckle. A hard or hot udder will be mastitis (inflammation of the udder) and will need immediate attention. Antibiotics and hand milking may save it.

Sore teats from over-suckling or from sharp teeth need antiseptic cream, regular stripping and protection from sucking; the lambs will need bottle feeding until the ewes have healed. Pustules on the teats and on ewe and lamb muzzles will be orf, a viral infection that is highly contagious to sheep and man; only time will heal it but take veterinary advice on how to manage it.

Prolapse of the uterus can occur after lambing. The uterus is pushed out through the vulva. Keep it clean before returning it or waiting for professional help.

FOSTERING

One job for the shepherd is to ensure that every ewe rears at least one lamb and that every lamb has a mother. There may be ewes without lambs because the lambs died. There may be lambs without ewes because the ewe died, had insufficient milk or had more lambs than she could rear (normally a ewe is expected to rear only two lambs). Any of these situations may mean that a lamb needs to be fostered onto a ewe.

In small flocks recipient ewes and foster lambs are rarely available at the same time; one answer is to take a lamb from a ewe with twins. Another answer is to link up with other flocks, but always be aware of the risk of introducing disease. In the UK 'lamb banks' are operated by local National Farmers' Union offices that put farmers who want lambs in touch with farmers with surplus lambs.

In theory foster and natural lambs should be matched for size but in practice any strong lamb that is capable of sucking will do. Fostering a small triplet onto a ewe with a large single – which is a common situation – may look comical, but lambs adopt their own teat so it is unlikely that a large lamb will take the smaller one's milk supply.

Check that a ewe has milk in both quarters before fostering on a second lamb. Do not foster a second lamb on to a ewe lamb and do not foster onto ewes which have had infectious abortion.

The principles of fostering are:

- Ewes are reluctant to accept a lamb from another ewe – so they need to be tricked.
- Ewes recognize their lambs by smell, especially around the tail and head – so the smell of foster lambs must be masked.

Methods of fostering include:

- Rubbing on.
- Skinning.
- Masking smells.
- Simulating birth.
- Adoption crates.
- Dogs.

Rubbing On

The most successful system is to 'rub on' a foster lamb while the ewe is lambing. The foster lamb is soaked in the ewe's amniotic fluids and given to her as soon as possible after the first birth. An older lamb should be restrained by having a front and diagonally opposite back leg tied together. This has the dual effect of making it look like a helpless new-born lamb and prevents it from gorging itself on colostrum. It may help to give it a good feed before joining the ewe.

Recipient ewes are those that are expected to end up with only one live lamb. If she subsequently has another lamb the foster lamb can be taken away and nothing will be lost.

Skinning

Where a ewe has a dead lamb it can be skinned (Fig 91) and put on the foster lamb. Skinning takes practice but is usually successful. A ewe waiting for a lamb should be penned with her dead one. This will stop her fretting and she will recognize the smell when the skin is put on a foster lamb. Skin it out of her sight.

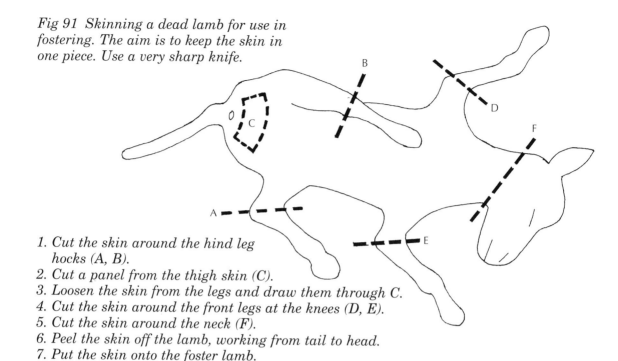

Fig 91 Skinning a dead lamb for use in fostering. The aim is to keep the skin in one piece. Use a very sharp knife.

1. Cut the skin around the hind leg hocks (A, B).
2. Cut a panel from the thigh skin (C).
3. Loosen the skin from the legs and draw them through C.
4. Cut the skin around the front legs at the knees (D, E).
5. Cut the skin around the neck (F).
6. Peel the skin off the lamb, working from tail to head.
7. Put the skin onto the foster lamb.

Leave it on for two to three days or until confident that the lamb has been accepted.

Masking Smells

Ewes identify their lamb by smell. If a lamb is being fostered on with a natural lamb the trick is to make them smell the same. The danger is that she might reject them both.

- Put both lambs in a dustbin with any afterbirth or foetal fluids. Some shepherds save foetal fluids from a ewe until a foster lamb is available. Put the dustbin in the ewe's pen for about an hour before releasing them. If there are no foetal fluids they may need to live there for a while and be taken out for suckling until the combination of binning, plus the smell of the ewe's own milk passing through both the lambs, make them smell the same.
- Proprietary sprays – or perfume – mask the smell of both lambs. This is often not successful and again there is risk of her rejecting both. A similar trick is to daub the heads and tails and her nose with molasses which she likes and licks off.
- Wash both lambs in warm salty water.
- Put a stockinette tube onto the natural lamb, cutting holes for the legs and making sure the neck and tail are well covered. After three hours remove it, turn it inside out and put it on the foster lamb. This system is reported to have been successful on foster lambs of up to twelve days old.

Simulating Birth

Put a ewe on her side and gently dilate the cervix with a lubricated finger for a minute to simulate birth and convince the ewe that she has had another lamb. She should be given the foster lamb, and her own lamb taken away for half an hour. This technique has been successful several days after ewes have lambed.

Adoption Crates

These hold the ewe by the neck and give her no choice but to suckle the lamb. They are invaluable in some flocks but should not be necessary in a small flock.

Dogs

Taking or tying a dog near the foster pen is thought to make the ewe protective towards her foster lamb and speed the bonding.

Anaesthetic

Sprays that anaesthetize the nasal receptors and stop the ewe smelling have been used with success.

If a ewe is suspicious of her foster lamb and reluctant to let it feed, restrain her regularly to let the lamb suck. When her milk has passed through the lamb she may recognize and accept the smell. Do not persist longer than about four days and do not leave a lamb with a ewe that might injure it.

Sometimes ewes that appear to dislike their lamb in the pen will become protective when turned out.

Hungry Lambs

Once it has had colostrum for twenty-four hours a hungry lamb can be topped up with ewe milk replacer (not cow's milk) from a bottle. Various teats are available to fit most domestic bottles but among the best are those with ball valves and that fit screw top bottles (Fig 74). For a quick and accurate mix find two measuring containers that give the correct proportion of milk powder and water.

Hungry lambs take a bottle quite readily – offer a finger first then substitute the teat. Feed little and often to keep them satisfied – around 50ml/kg body weight every three hours.

Candidates for topping up are any hungry lambs. Identify the reason for the hunger, such as a large litter, poor milk yield or sore teats. If fostering is not an option leave the family together and continue to top them up. They learn to come for their bottle very readily and topping up may only be necessary for a few days. Where more than one lamb is being topped up it may be easier to put them on an artificial rearing system (see Chapter 10).

Castration and Tailing

Ram lambs which are destined for slaughter need not be castrated unless they are likely to be kept after five months of age, when they mature sexually and become a management problem. The advantage of leaving them entire is that they are leaner and grow 7 per cent faster than castrates. However, there may be resistance to entire ram lambs in some markets. The rubber ring method (Fig 93) is widely used in small flocks. A bloodless castrator which crushes the spermatic cord is used on older lambs.

Lambs are tailed (Fig 92) for hygiene reasons, and those kept for breeding will benefit as adults from not having long tails. Slaughter lambs need not be tailed unless they are destined for the store and heavy hogget market when they may get dirty during the finishing period. Very short tails which are banned in some countries, including the UK, may predispose prolapses in breeding ewes.

The End of Lambing

When lambing has finished, clean the housing and equipment with a broad spectrum disinfectant available from agricultural merchants. Remove bedding and soak the floors and walls, scrub troughs and hurdles and put in the sun to kill germs. Clean all equipment including water buckets. Do any repairs and make a note of any changes or improvements for next year.

Fig 93 (Left) *Rings for castration should not be used after seven days of age and nor should lambs be castrated within a day of birth otherwise it will interrupt colostrum intake. The rings cut off the blood supply to the scrotum which drops off in 10–20 days. A veterinary surgeon or experienced shepherd will demonstrate the correct procedure. Do not open the jaws of the applicator too wide because it weakens the rubber ring, make sure the testicles are in the scrotum and do not apply it too high where it traps the teats or urethra. Slow tissue death leaves a wound for infection to enter; if possible, treat the wound with an antibiotic spray.*

Fig 92 (Right) *Tailing with rubber rings. Enough tail must be left to cover the vulva of the ewe and the anus of the ram. Putting the ring just below the base of the under-tail fleshy area is a standard guide, but err on the side of caution because some lambs can outgrow their tails and leave everyone embarrassed. In the UK they should not be used on lambs of more than seven days of age.*

No ID	Date lambed	Litter	Sexes	Sire	Weight	Lamb deaths			Foster	Special treatment		CS	Comments
						Age	*Sex*	*Reason*		*Ewe*	*Lamb*		
24	1/3	2	R/E	Suff	4.5/4kg	2hr	E	Lain on 22	R from	None	None	3	Assisted breech

Table 7 Ewe Lambing Records

Analysis of Lambing

No. put to tup	25	Reasons for death:	Watery mouth	1
No. lambed	22		Accident	1
% lambed	88		Stillborn	2
% lambing to first service	80	% of lambs dying		7
% lambing to 2nd service	15	No. ewes assisted		3
% lambing to 3rd service	5	% ewes assisted		14
No. not lambing	3	No. lambs assisted		3
% not lambed	12	% lambs assisted		7
Reasons for not lambing: Aborted	0	Reasons for assistance:	Large	1
Barren	1		Breech	2
Died	1	% lambs born per ewes lambing		200
Culled	1			
Reasons for deaths: Pregnancy toxaemia	1	% lambs reared/ewes tupped		164

Litter sizes:

No. lambs born alive	44	Sets singles	6	% singles	13.5
No. stillborn	2	Sets twins	17	% twins	77.0
Reasons for stillbirths Dystokia	2	Sets triplets	2	% triplets	9.5
No. died in 48hr	2	Sets quads	0	% quads	0
Reasons: Watery mouth	1	Others	0		0
Lain on	1	No. ram lambs	26	% ram lambs	59
No. died 48hr to 2 weeks	1	No. ewe lambs	18	% ewe lambs	41
Reasons: Fell in trough	1	No. lambed ewes not rearing a lamb		1
No. died 2 weeks to weaning	0			
Reasons:	–	Reasons: No milk/mastitis		
Total no. deaths/stillborn	5	No. lambs fostered	3	
Reasons for deaths: Starved	0	No. lambs hand reared	0	
Lain on	1	A separate record can be kept for ewe lambs		
Foxes	0			

10 Lamb Rearing

After lambing, when they have mothered on and the weather is suitable, ewes and lambs are put onto sheltered grass or forage crops.

A combination of rain and wind causes windchill (loss of body heat) and the effect on young lambs should never be underestimated. Shelter can be provided by:

- Hedges.
- Windbreak material (Fig 94).
- Lamb jackets (Fig 95).
- Bales (Fig 96).

Lamb jackets offer good protection for the first few days, but choose a secure design otherwise they will litter the fields, and fit them while the lambs are in the mothering-

on pen to make sure the ewe does not reject a lamb in a jacket.

Before turning out:

- Make sure that ewe and lambs are easily identified.
- If foxes are known to be a risk take steps to protect the lambs. Deterrents include electric fencing, sheep bells (or bells sold for parrots) around the necks of 20 per cent of ewes, ferret bells around the necks of lambs, a daub of tar on the back of lambs' necks, yellow flashing lights around the perimeter of the field, or bright orange lamb jackets.
- Put stones in low water troughs to prevent lambs drowning. Where the trough

Fig 94 For sheep of any age a zigzag of plastic windbreak material will give protection from all weathers in all directions.

is high, keep a bucket of water alongside for the lambs but make sure that the level in the bucket does not fall so low that the lambs cannot reach the water over the rim.

After turnout:

- Check regularly for mismothering or rejection – especially in bad weather when the flock congregates around shelter. During the first few days some lambs may look as if they have been abandoned. There is no need to panic. The ewes choose a place to leave lambs while they go off grazing and return to them regularly for suckling. Later on the roles will reverse and the lambs will seek out the ewes.
- Watch for lambs whose tails are stuck down over the anus with faeces.
- Look out for hungry lambs. They usually hang around their mother instead of going off to play or sleep. They walk

Fig 95 Jackets protect baby lambs from windchill. Orange jackets are said to deter foxes.

backwards, sucking as the ewe walks forward to graze, or pinch milk from between the back legs. Check the ewe for sore teats or mastitis; if she is simply a

Fig 96 A simple arrangement of small bales gives shelter to lambs whatever the direction of the weather but make sure they cannot fall down the gap in the middle.

137

poor milker the lambs may need to be topped up. Sometimes milk supply improves with time and the lambs eventually ignore the bottle.
- Watch for hypothermia – full-fleeced ewes are unlikely to seek shelter and their lambs may be exposed.
- Watch for hypomagnesaemia.
- Allow plenty of trough space to avoid trampling or mismothering lambs and feed before nightfall to allow enough daylight time to sort themselves out.
- If temporary shelter is used, move it regularly to avoid a build-up of disease and mud.

Health

Once lambs are two weeks old the risks to their health diminish. However, losses from post-lambing to sale can average 3 per cent and problems to watch for include:

- Starvation and hypothermia.
- Tailing or castration wounds when the tail or scrotum falls off. Treat with an antibiotic spray as a precaution.
- Lameness due to scald (inflammation between the digits), especially in damp weather and long grass. Treat with an antibiotic spray.
- Coccidiosis, caused by a parasite picked up from around feed troughs especially in intensive systems. Lambs are seen tucked up with mild diarrhoea.
- Coccidiostats can be added to the feeds of both the ewes and the lambs.
- Nematodirus.
- Clostridial diseases. Vaccinate at ten to twelve weeks when colostrum antibody protection has worn off. Usually only necessary for lambs that are retained in the flock or not destined for slaughter during the summer.
- Flystrike/headfly.
- Pine, white muscle disease and swayback are the results of cobalt, vitamin E and copper deficiencies respectively and

are manifest by unthrifty and unsteady lambs.
- Orf.
- Urinary calculi (urolithiasis), associated with intensively fed castrated ram lambs. Crystals form in the urine and block the urethra and may cause the bladder to rupture.

Problems to watch for in ewes include hypomagnesaemia, mastitis, footrot, orf and flystrike.

MILK PRODUCTION

Lambs are totally dependent on milk during their first four weeks so priority is given to feeding the ewe to maximize yield (see Chapter 5). Feed concentrates that provide 12.5 ME and 18 per cent protein from quality raw materials. A protein deficiency at this stage can trigger a dramatic fall in milk yield within three days. When the spring grass is 6–7cm high, concentrate feeding can be reduced or stopped – especially in ewes rearing singles.

Fig 97 shows a typical lactation curve for a ewe. Milk production peaks at three to four weeks and feeding in early lactation can increase total production by both raising and prolonging this peak.

Lamb growth rate is a good indication of a ewe's milk yield. Multiplying the daily growth rate for the first four to six weeks of a single lamb by five (or the mean growth rate of twins by ten) will give the daily milk production within 90 per cent accuracy. For example, a ewe with a single lamb whose growth rate averages 450g per day is giving $450 \times 5 = 2.25$ litres. Conversely, a ewe would need to give 2.5 litres of milk per day for her twins to achieve a 250g daily growth rate.

After a lamb is six weeks old, grass gradually replaces milk in its diet but, because it takes 3.5–5.0g of grass dry matter to replace 1g of milk, the growth rate continues to be influenced by milk right up to weaning.

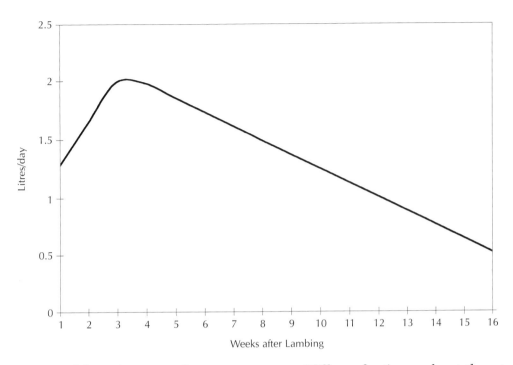

Fig 97 A typical lactation curve for an average ewe. Milk production peaks at three to four weeks after lambing. Good feeding before and immediately after lambing should produce a high peak and slow its decline. After six weeks a lamb is deriving a substantial proportion of its food from pasture and is less dependent on milk, so there is little point in continuing to feed concentrates to the ewe. This curve also illustrates why leaving lambs on their mothers after fourteen weeks is of no benefit.

Ewes suckling twins give about 40 per cent more milk than similar ewes with singles and reach peak lactation earlier. The decline in milk production is faster in the ewes with twins, and yields are about the same after twelve weeks. Ewes rearing triplets will produce 10 per cent more milk than those with twins. First-lambers usually give 75 per cent of their potential adult yield.

THE GROWING LAMB

Understanding the growth rate and growth pattern of lambs is important when planning their feeding and marketing.

Lambs grow fast and convert their food very efficiently during their first few weeks of life; they can double their birth weight in two weeks. When they reach puberty – at about half their adult weight – growth rate slows progressively to maturity (Fig 98). This means that it is effective to feed them well when they are young.

Food conversion ratio (FCR) is high in a young lamb. This is the measure of the amount of food needed to produce the same amount of liveweight, and the figure is often quoted in literature from feed manufacturers. A lamb eating a compound feed can convert 3.5kg of feed to 1kg of bodyweight – giving a high FCR of 3.5:1.

From birth to weaning the average daily liveweight gain (DLWG) of a lamb in a commercial flock is around 0.2–0.3 kg or 1.5–2kg per week, although gains in excess of 4kg per week are possible.

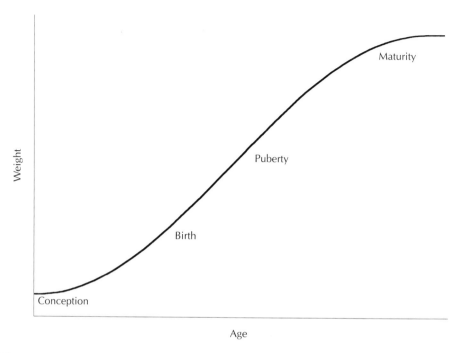

Fig 98 The growth curve of sheep shows why lambs are usually slaughtered at around 50 per cent mature weight (puberty) before their growth rate slows down and before the percentage of fat in the carcass increases substantially (Fig 99)

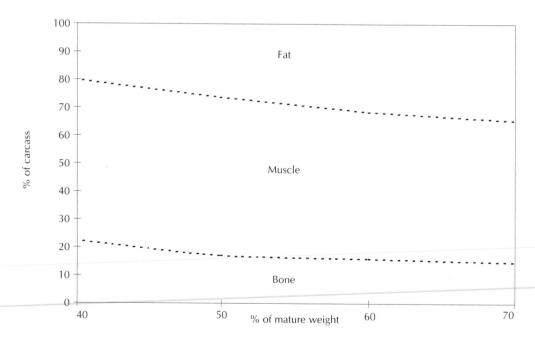

Fig 99 The growth pattern of lambs. As lambs get older the proportion of bone and muscle in their bodies decreases as they lay down an increasing proportion of fat.

140

In early lactation a single lamb will grow almost twice as fast as a twin lamb. A single can put on 400–500g per day on milk and grass and be finished at ten weeks. The difference narrows as they grow older and depend more on grass, but twins rarely catch up. Early born singles can cause marketing problems in a small flock when they become heavy and fat waiting to make up a group for selling. The temptation is to keep them as replacements but this is not wise if selecting for prolificacy.

There is a strong relationship between birth weight and weaning weight with every additional 1kg at birth resulting in an estimated 2–3kg extra at weaning. The age of the ewe is also a factor in growth, with young ewes having slower growing lambs; ewes of around five years of age will produce their heaviest lambs at weaning.

Growth Pattern

The natural growth pattern for a lamb is bone first, then muscle, then fat. As lambs get older (Fig 99) their percentage of fat progressively increases to around 30 per cent of the carcass weight. Because the skeleton is early maturing and the fat late maturing, some growing lambs can lose their chubby look for a while and become rangy – giving the worrying impression that they are not thriving.

As a rule small, light breeds are early maturing. This means that they complete the growth process of bone, muscle and fat quickly and lay down fat at an early age. Conversely, heavy breeds are late maturing and grow more bone and muscle before getting fat. They can be sold for slaughter at a later age than the light breeds. Breeding and selection programmes within some breeds have modified this trait but, generally, quick maturing breeds are suited to early lamb production and large slow maturing breeds are chosen for late lamb and store lamb production.

FEEDING LAMBS

Grass is the best and cheapest feed for sheep, and lambs can be reared on it without being fed concentrates. However, in the real world, lambs will often need some creep feed. Creep feeding is the practice of giving concentrates to lambs while they are still sucking.

Lambs are creep fed when:

- They need fast growth for the early market or shows.
- Triplets are being reared by their dam.
- There is a shortage of grass.
- They are reared artificially.

Creep feed is commercially available as small pellets and is fed where it is accessible to lambs but not to ewes (Figs 100 and 101) Intensive creep feeding can give an average growth rate of 400g per day and hastens the development of the rumen to digest roughage. The need to creep feed depends on how quickly the lambs need to grow and how much pasture is available. Sucking lambs which are on good pasture may not need extra feed.

Creep feed must be formulated specifically for lambs. They should never be given feed or minerals that are formulated for ewes; urolithiasis is one of the risks of this practice. It should be offered *ad lib* from ten days of age but lambs eat very little until three weeks. Initially offer only a taster and make sure that it is always fresh, dry, clean and available. Birds are a menace, not only eating the food but also fouling it.

Depending on the reasons for starting creep feeding it is best to continue through to sale or weaning, or until grass or forage crops are adequate, to prevent a check when it is withdrawn.

Table 8 shows typical home-mixed rations, but for small flocks it is probably cheaper to buy ready-mixed. Consumption depends on the milk and grass supply but it is wise not to buy too much at a time especially towards

Fig 100 Some creep feeders allow the lamb to get its head into a trough but the gaps are too narrow for the ewe. This feeder has a hopper for holding several bags of feed and it allows lambs to feed from either side. The space widths are adjustable.

Fig 101 Some feeders allow the lambs into a shelter which incorporates a feed trough. This was made from an old water tank turned on its side. The feed trough is protected from the weather behind the windbreak plastic gate and a simple creep gate allows in only the lambs. Any fenced-off area, or a pen made from hurdles, will make a creep feed area provided the barrier keeps out the ewes. The system should be portable to avoid a build-up of mud and disease and to accompany the flock when it is moved. Encourage the lambs to investigate and use the feeder by siting it near the ewe feeding area; and a few bales close by attract lambs to play.

Table 8 Simple Home-mixed Creep Feed

To make 100kg

Rolled barley	77.5kg
Soya bean meal	20.0kg
Mineral/vitamins (lamb)	2.5kg
or:	
Rolled barley	60kg
Molassed sugar beet feed	20kg
Protein pellets	20kg

the end of the feeding period. Lambs can eat around 1kg per head per day of creep feed by eight weeks of age but once fresh grass is available they quickly lose interest in concentrates.

WEANING

The process of converting grass into milk and then milk into meat is biologically inefficient at the tail-end of lactation and traditionally lambs are weaned at around 12–14 weeks when, as Fig 97 shows, milk supply is minimal. However, if slaughter lambs are very close to finishing at fourteen weeks it may pay to delay weaning and sell them off their mothers to avoid a weaning check; and in small flocks, weaning later so that lambs can be sold off their mothers while they are still suckling may be more convenient where there is not enough land to keep the weaned ewes and lambs apart. Lambs being reared by ewe lambs should be weaned no later than twelve weeks.

Weaning too late risks the problems of:

- Ewes and lambs competing for grass.
- Entire ram lambs of four to five months old serving their own dams.
- Ewes having insufficient time to recover for tupping.

Weaned lambs should be wormed and stocked on clean, leafy, clovery 5–8cm pasture and when they have had the pick of the grass the dry ewes can tidy up behind them.

At weaning, take the ewes away and leave the lambs in a secure and familiar field. Ewes are easier to move from the weaning pen; lambs have not learned to flock and are difficult to drive.

Both flocks should be out of earshot but this is not easy to arrange on a small property. Housing the ewes helps to muffle their calls. Normally ewes and lambs call for about two days then they settle, but if given the chance they will mother back on again until at least three weeks after weaning.

It is better to wean the whole flock at one time, even if there are tail-end lambs, because the ewes must have time to recover and because it is easier to manage the ewes as one flock. Ewes can go on bare pasture or be housed on straw (always with water) to encourage quick drying off. Watch for mastitis and physical damage as they fill up. The full udders may look uncomfortable but trying to ease them by hand-milking will stimulate production and risk introducing disease by keeping the teat orifice open (see Appendix I, Fig 116). Where mastitis is a problem after weaning, some veterinary surgeons recommend dry ewe therapy in which an antibiotic is inserted into each teat canal.

Early Weaning

Hand-reared lambs are usually weaned at around six weeks of age onto an intensive rearing system. (see Artificial Rearing box) but early weaning is also appropriate in the following situations:

- In housed flocks – where lambs are weaned at 6–8 weeks, the ewes are turned out and lambs finished indoors.
- In a frequent lambing system (three times in two years) when ewes need time to recover.
- When there is a shortage of grass.
- For ewes with triplets.
- In hill flocks when twins are 'split' and one is taken away for finishing and the other left with its mother.

Slow-growing Lambs

Lambs which grow more slowly than expected are a problem in any flock. They are usually the result of a shortage of feed caused by overstocking or drought, but may be the result of worms or a trace element shortage such as cobalt. The outcome is that they miss their market and the longer they hang around the more they become prone to worms and other diseases, compete with ewes for grass for flushing, have accidents and incur extra costs and cash flow problems. In the small flock, where space is a problem, they simply become a nuisance. More so if the ram lambs were not castrated.

In following years it may mean improving nutrition by changing the lambing date to match grass supplies or improving grassland or crops to match sheep demand.

In the meantime there are a number of ways to cope with them:

- Sell as store lambs – probably through the livestock market.
- House or confine and finish on concentrates. Shearing them may stimulate faster growth.
- Store on a maintenance diet of grass or hay and finish them later (see Store Lambs later).

Feeding for the Market

The way that lambs are fed and reared will depend on when they are born and the market for which they are intended.

Autumn-born lambs from Dorset Horn flocks and crosses, wintered on forage crops or grass, may not need extra feed to get them finished at around sixteen weeks for the early market. Decide the optimum weight and selling date and weigh them regularly, giving creep feed if necessary to keep on target. Like store lamb finishing, this system relies on controlling growth rather than hastening it; lambs must not be finished before the market wants them.

Early lambs, expected to finish at 10–14 weeks, may need some creep feed to grow at around 2.5kg per week to hit the early high prices (Fig 103). Some lambs are sold finished off their mothers (FOM) before weaning.

Spring lambs are weaned at fourteen weeks and then finished over the next few weeks on pasture or forage crops. They need to grow at around 1.8kg per week before weaning and 1.4kg per week from weaning to slaughter, which can be achieved on aftermaths and late summer grass and should not need concentrate feed.

Late lambs are weaned at fourteen weeks and, if prices are poor, kept or sold as store lambs for finishing during the autumn and winter. Concentrate feeding is probably only necessary for quick finishing in the winter.

STORE LAMBS

Buying and selling store lambs (see Chapter 1) is a specialist sector of sheep farming and is not particularly suited to the small flock other than as an introduction to keeping sheep. Storing on a restricted diet, followed by an improved diet, results in compensatory growth when sheep grow faster than would be expected on that diet.

There are 'short-keep' store lambs and 'long-keep' store lambs. As a general rule short-keep stores are bought within 7kg of their slaughter weight (see Chapter 11). Many will finish within 6–8 weeks on autumn grass at 20–30 lambs per hectare, gaining weight at 1kg per week and sold at the end of the season (October to January).

Long-keep stores are bought within 10–20kg of their slaughter weight and may be kept for up to five months. They are usually grown slowly (about 0.2kg per week) on a low-energy diet such as winter grass so as to grow a large frame and are finished in the final 4–6 weeks on forage crops and concentrates for the hogget trade in the spring. They must be kept off new sheep leys to avoid parasitic infestation for next year's lamb crop and must be off next year's grazing land by the end of December.

Autumn grass and forage crops give weekly growth rates of around 0.75–1kg; a hectare of rape can graze sixty lambs for six weeks and a hectare of swedes will feed sixty lambs for eighteen weeks. Growth on roots tends to be slower than on forage crops, and roots need protein supplementation. There is normally adequate protein in leafy forage crops (rape and kale). Lambs may begin to lose their milk teeth around nine months of age which can make it difficult for them to eat roots.

Waste is high in most forage crops that are grazed *in situ* although some producers claim that there is less trampling with set stocking than strip grazing. Lambs must have a grass run-back or dry lying area so that they do not have to lie in muddy fields, and be introduced to forage crops gradually over 7–10 days. Brassicas such as kale and rape can induce iodine deficiency so may need rationing, and copper deficiency can be induced in lambs on root crops if they ingest soil. Always offer hay when sheep are on these low dry matter crops.

Store lambs can be shorn in late summer in order to keep them clean, prevent them from tangling in hedges, increase their appetite so that they grow faster, and take less space if housed.

Artificial Rearing

Lambs that do not have a mother, lambs bought in from other farms and lambs from dairy ewes can be reared successfully on artificial milk (ewe milk replacer).

Avoid rearing one lamb on its own – they are sociable creatures and need playmates. Leave a lamb with its mother, where appropriate, and top up with a bottle until it has a companion. Ideally lambs should be reared in groups of no more than eight and matched for age and size.

Ewe milk replacer powder is normally sold in 10kg bags – enough to rear one lamb. Mix and feed it accurately and according to the instructions. If the mix is too weak the lambs do not thrive and if too strong it can cause dehydration. The quantity to feed will be specified in the instructions, but as a rule of thumb feed about 50ml/kg bodyweight in four well-spaced feeds a day.

The lambs are penned on straw in a draught-free area with access to fresh water. Warm ewe milk replacer is fed by bottle for 3–4 days by which time they should be strong enough to use a feeding system.

With small numbers it is probably easier to continue to bottle feed three times a day until weaning. Larger numbers can use a cold milk feeder (Fig 102). Lambs are content on *ad lib.* feeding systems where the 'little and often' drinking pattern is natural; the cold milk deters them from gorging.

Feeders can be built from teats, tubes and non-return valves available from agricultural merchants. Introduce them to the feeder when they are clamouring for their next feed by filling it with warm milk and holding them to a full teat which has milk smeared on the end. Once they get proficient the milk can be fed cold. The feeder can be topped up twice a day with half their daily ration, and to make sure that all the lambs have the opportunity to drink, it should never be empty – other than for regular cleaning.

Lambs must always have access to clean water and, after a week, have hay and fresh creep feed. Water and feed containers should be supported off the ground so that they cannot be walked in or slept on. Keep pens and feeding equipment clean and move feeders around the pen to prevent a build-up of urine and dung.

The lambs can be weaned at 5–6 weeks of age, weighing around 15kg and when eating 200–250g per day of creep feed. Most milk manufacturers advise weaning abruptly. Slow weaning accustoms the lamb to eating less and they stop growing, whereas abrupt weaning leaves the lamb hungry and it eats something else.

Lambs can be finished on compound feed or put on worm-free, quality grass. Hand-reared lambs suffer a growth check when they first go onto grass and are very susceptible to worms.

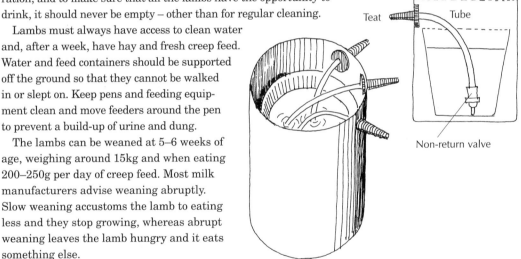

Fig 102 For ad lib. *cold milk feeding or for four-times-a-day warm milk feeding a row of teats allows lambs to drink from a bucket via a tube and non-return valve. Once they are sucking strongly the valve is not necessary. Teats can be incorporated in a large container with a milk bucket inside, as shown. There are numerous systems and arrangements which can be customized to suit the flock.*

Basic timetable for artificial rearing

Day 1	Feed colostrum.
Days 2–4	Bottle feed warm milk replacer.
Day 5	Introduce to the feeder.
Day 7	Introduce hay and creep feed.
Day 42	Wean.

11 Marketing and Business

There are several points to consider when marketing sheep and sheep products:

- Identify the market before going into production.
- Find out from the buyer exactly what he wants and when he wants it.
- Make sure the buyer knows at the outset the quantities and quality you are likely to have.

Scale is one of the disadvantages of small flock production. Small producers have very little clout when it comes to both buying and selling. It may help to either get together for selling, join an existing co-operative marketing group or simply set up a meeting point and share transport to markets, sales or abattoirs.

SELLING

Most livestock can be sold through livestock markets or advertised in local papers. Whatever the system make sure that details of breed, age, status (for example, in-lamb), purpose (for example, fine wool), management (for example, organic) and reasons for sale are made known to buyers. Make sure the auctioneer knows of any health schemes, vaccination programmes, wormers and any relevant details of their breeding. Match in groups according to type, breed, age or teeth. One odd one will spoil the pen. Stock going to market must be healthy. If sick, lame or diseased sheep are unloaded at an abattoir or market they

may be destroyed on the spot and the owner fined.

Pedigree breeding stock can go to specialist breed shows and sales either locally or nationally or be advertised and sold privately. Flock and pedigree records should be available. Breed societies will be the source of information on suitable sales. Stock will need to be good examples of their breed and have all the attributes of good healthy breeding animals. They should be well fleshed but not fat; feeding to get them right on the day is both an art and a science, but over-feeding young stock can reduce their potential for breeding. They may need to be halter trained, washed and trimmed and the shepherd must also look tidy and well presented.

Cull ewes are sold through most livestock markets and direct to abattoirs. Some abattoirs offer an on-the-spot price for the live animals but most pay on carcass weight after slaughter. Most culls go for processing or to ethnic groups who eat mutton. Fattening thin ewes or holding culls until the market improves may be worthwhile, but few producers market old ewes with the same care as lambs. The trend is to get them off the property as soon as possible especially where land is limited. Mutton is often not readily available through butcher's shops and supermarkets so there may be a private or local niche market.

Store lambs are sold from the autumn onwards usually through the livestock market and should be well grown, clean, healthy and presented in even-sized batches. Vaccination programmes and wormers should be reported to the auctioneer who should announce them when the sheep are sold. The same applies to ewe lambs sold for breeding except that they should have a better finish so that the buyer can mate them successfully.

Organic meat has a specialized market and anyone who has gone through the process of registering land for organic production should know the outlets. These are likely to be local butchers or private sales. There may be restrictions on the distance organic lambs can be transported and restrictions on which abattoirs can kill organic lamb.

Wool marketing is discussed in Chapter 7.

Milk products are likely to be sold to local retail outlets, usually delicatessens, or the milk may be frozen and sold to a manufacturer for processing into cheeses and yoghurts.

Slaughter Lambs

Most sheep flocks will sell some or all of their lambs for slaughter. Even pedigree breeders will sell a large proportion for slaughter when they do not attain breed standards. There tends to be a pattern to lamb sales and prices (Fig 103).

Slaughter lambs have a number of outlets but fundamentally are sold liveweight through livestock markets or deadweight direct to abattoirs and butchers.

Liveweight sales are primarily through local weekly livestock markets and prices

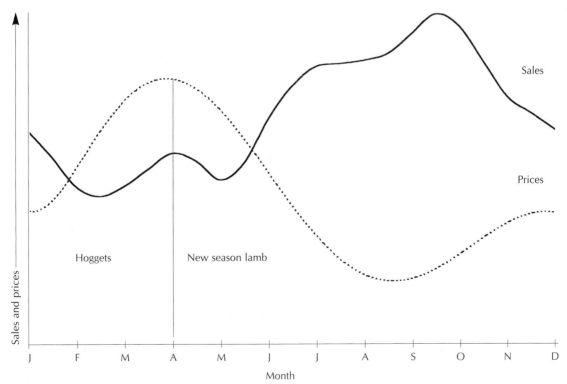

Fig 103 The normal pattern of lamb prices and sales in Northern Europe. Prices peak at the traditional Easter period. Sales peak in the autumn when lambs are finished off grass. Unfinished lambs are kept as stores and sold during late winter and early spring as hoggets to fill the gap before new season lamb becomes available. Cull ewe prices tend to follow the same pattern as lambs.

are expressed in pence per kilogram of liveweight (p/kglw). Livestock market prices tend to establish the prices for the whole industry. The advantages of the system are:

- It is convenient for small batches at irregular intervals.
- If you do not like the price they can come home again.
- It is a good place to chat about the sheep industry and have a day out.
- Within reason poorer stock can be 'dumped' there. There will almost always be a buyer, at a price.
- Because it is competitive, the vendor feels that the price is fair.
- The markets are held regularly so marketing is easy to plan.

On the debit side there is normally more transporting and hassle involved for the sheep because they are unloaded at market, penned for much of the day and then reloaded and transported for slaughter.

In the deadweight system prices are paid in pence per kilogram of carcass weight or dead weight (dw) – not including the offal, pelt and fleece which the abattoir will keep. The advantages are:

- Carcass data from the abattoir provides information on the quality of the lambs.
- The producer usually knows in advance the prices being paid.
- Transport is kept to a minimum.
- Information on disease problems, such as liver damage, may be available.

On the debit side there can be uncertainty about when the buyer will take the lambs. If a carcass is rejected or marked down because it is not up to the standard that the buyer wants the vendor has to accept a lower price. Some abattoirs will return rejected carcasses to the producer; a freezer full of over-fat lamb is a salutary lesson.

Electronic marketing is described as offering the best of both worlds. Lambs are select-ed on the farm and sold live via an electronic auctioning system and are later transported direct to an abattoir. For small groups of lambs this is probably too complicated.

Having lambs slaughtered and butchered for private sales is a good scheme for owners of small flocks who find a niche market for rare breeds, organic lamb or just locally grown lamb. Local restaurants may be customers but probably want a regular supply. Local markets may also exist for mutton. Private sales may be subject to hygiene and marketing laws to avoid meat going unlawfully into the food chain and ensure that carcasses are fully inspected at the abattoir. Anyone considering private sales of meat should investigate the current regulations.

Mobile slaughtermen operate in some areas and are popular with producers who do not want to stress their animals by taking them to an abattoir.

THE PRODUCT

There is little point in growing a product if it is not what the market wants. What the meat buyer wants is:

- An optimum carcass weight.
- Minimum fat.
- Good killing-out percentage.
- A high percentage of valuable cuts.
- A clean, undamaged carcass with no faults.

What the consumer wants is:

- Good value.
- Minimum fat.
- A belief that the lamb was reared under good conditions of welfare.
- That the product is safe to eat.

All this may seem irrelevant to the small flock but the industry depends on the consumer buying lamb; if it is not to their liking they will buy something else. So every producer, however small, should consider the consumer.

Carcass Weight

The ideal carcass weight in the UK is 16–20kg (about half the liveweight) although carcasses as light as 8–13kg – usually from hill flocks – find a market in southern Europe. There are outlets for heavier lambs in the processing and catering trades but big lambs produce big joints which cost more than the housewife is prepared to pay.

Lighter carcasses often attract a higher price per kilogram than the heavier ones but, as long as they are not fat, the heavier ones often earn the most money per head.

Fat

Modern sheep breeds have fat under the skin (subcutaneous) and within the muscles (marbling). Subcutaneous fat tends to be laid down before internal fat which makes it a good guide when selecting lambs for slaughter (*see* later).

Primitive and rare breeds tend to lay down fat internally around the organs and have lean flesh.

Subcutaneous fat is an aid to cooking but too much fat is a waste because most consumers do not want it; the producer of fat lambs pays a double penalty because it costs feed to produce it, and then he is paid less money for the lambs.

Killing-out Percentage

The killing-out percentage (KO percentage) is the percentage of the live lamb finishing up as meat, fat and bone (carcass) after slaughter and after the fleece and offal have been removed. The killing-out percentage of a modern breed of young lamb is usually around 45–50 and for older lambs (hoggets) it is 40–45.

Where lambs are sold deadweight to a butcher the KO percentage is important because the producer is paid on carcass weight. Where lambs are sold through a live market the buyers estimate the KO percentage and bid appropriately. It is useful to have an idea of this figure by comparing carcass weights from the abattoir with the liveweights of the animal before slaughter – bearing in mind that it is affected by gut fill, fleece weight and the loss of weight (around 1kg) during transport.

Value Cuts

Distribution of lean meat is important because various cuts (Fig 104) have different values. Hind legs and loin represent about 40 per cent of the value of the carcass, and loin chops may be worth three times as much as breast.

Generally there is a relationship between conformation (body shape) and high value cuts. It is thought that a lamb with good conformation – blocky with a long, wide back, well-fleshed rump, fine head and short legs – is likely to have a good KO percentage and a good percentage of valuable cuts. The depth and thickness of the top of the hind leg is a good indication of overall muscling. Sometimes, however, good conformation may simply be due to fat.

The size of the eye muscle (*see* Appendix II) is an important economic trait because it gives a meatier chop. Breeders of terminal sires select for it.

Faults

The main faults found in slaughter lambs are:

- Bruises.
- Abscesses.
- Dirty sheep (which contaminate the carcass).
- Taint.

Young lambs have soft skins and bruise easily, leaving areas of carcass that must be

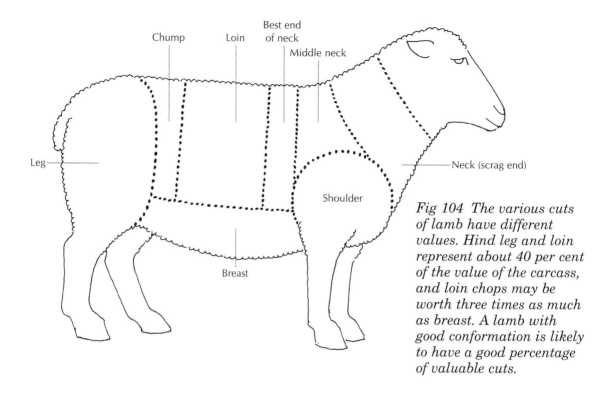

Fig 104 *The various cuts of lamb have different values. Hind leg and loin represent about 40 per cent of the value of the carcass, and loin chops may be worth three times as much as breast. A lamb with good conformation is likely to have a good percentage of valuable cuts.*

trimmed off, and the producer is penalized by a deduction from the price he is paid. Even pressing their backs while assessing fatness can leave a mark. Lambs should never be caught, handled or picked up by their fleeces, and penning and loading should be done with the minimum of crushing – avoiding all projections and snappy dogs. Take care that lambs do not get caught in brambles or barbed wire.

The value of the skin or pelt is not paid to the individual producer but is part of the meat trader's profit and influences the price paid for lambs. Damage from barbed wire, injection abscesses and skin diseases all affect the quality and should be avoided.

Wet and dirty sheep will be turned away from markets and abattoirs. Keeping them clean is important and is a matter of good management – avoiding scouring, muddy fields and gateways. Older lambs with long fleeces on forage crops are most at risk and dirt around the tail and belly should be

shorn or hand trimmed. Keeping lambs dry is a problem; even housing them will not guarantee a dry fleece.

Taint is a subjective issue and some abattoirs refuse to take entire ram lambs because of a fear of odour. Tainted meat in ram lambs has never been conclusively proven and any taint may have as much to do with feed as with sex. Flock owners who do not castrate rams should tell their buyer, especially if the lambs are more than five months old.

Welfare

There are a number of marketing schemes set up by farmers, supermarkets and welfare organizations which pledge that their lamb was produced in a welfare-friendly naturally-reared system. Some butchers and abattoirs pay a premium for such lamb. In reality most lamb is reared naturally and without cruelty.

SELECTION

Selection for slaughter is a compromise between weight and fat cover, but fat must be the major criterion; a lamb should be sold as soon as it has ideal fat cover even if it is lighter than planned.

As a rule of thumb a lamb reaches slaughter weight at around half its mature weight – mature weight being half the combined average adult weights of the dam and sire breeds. For example: a Suffolk ram of 91kg mature weight over a Cheviot ewe of 65kg would produce a lamb with a potential adult weight of 78kg, or an optimum slaughter weight of 39kg, to give a carcass weight of around 18–19kg at fat class 3 (*see* later) and a KO percentage of 48.

Generally a lamb that is slaughtered 10 per cent below the predicted half adult weight will be fat class 2 and those that are 10 per cent over the weight will be fat class 4. Ewe lambs get fatter than wethers (castrated ram lambs) so to achieve the same level of fatness wethers should be slaughtered at 5 per cent above the average slaughter weight and ewes 10 per cent below. In other words, be prepared to sell ewe lambs for slaughter a bit earlier and a bit lighter than expected. Entire rams and hoggets usually go to heavier weights within the ideal fat class; intensively reared concentrate-fed lambs finish at lighter weights and must be checked for finish twice a week during the final stages of feeding.

Classification

The market requirements for lamb are usually described in terms of carcass weight, fatness and conformation. This classification is a language that allows buyers to tell producers what they want and producers to produce it.

The language, represented by a grid (Fig 105), has twenty-nine categories of classification. So a buyer might say that his preferred carcasses are 3LU 16–18kg carcass weight – that is a lamb in fat class 3L with conformation U. Some buyers pay extra for their ideal carcass and pay much less for carcasses outside their specifications, so knowing their requirements is vital.

At the moment fat levels and conformation are assessed subjectively, but one day in the future probes and scans may assess them objectively.

Judging Fat

The amount of fat on a lamb is judged and described on the same basis as condition scoring ewes (Appendix 2) but expressed as fat classes 1–5. There is no muscle over the spinal vertebra – only skin and fat – so the back and the dock of the tail are the best areas for judging fatness along with the shoulder and the ribs (Fig 106). Visual assessment is no good – the lamb must be handled (Fig 107).

Livestock markets, marketing groups and other organizations often run practical courses on lamb selection at the start of the main marketing season. These are especially useful for the small flock owner with limited opportunity to practise.

Producers who sell deadweight can check their opinion with that of the abattoir after the animal has been slaughtered. Most abattoirs can identify individually tagged lambs, but where live lamb and carcass cannot be matched the data for the whole group will give an idea of whether lambs are being drawn too fat, too thin, too heavy or too light.

Selecting for Market

Lambs should be weighed regularly from around eight weeks of age to monitor and record growth rates, to adjust feed and to estimate the time of the first sales. Once they begin to reach slaughter weight they should be weighed and handled weekly.

A weigher suitable for ewes as well as lambs is a crucial tool on the sheep farm.

Fig 105 The classification grid, devised by the UK Meat and Livestock Commission, tells the producer what the buyer wants in a language they both understand. The fat class is based on the assessment of subcutaneous fat (Appendix 2) on a scale of 1–5 (very lean to very fat) with categories 3 and 4 subdivided into L (low) and H (high). As the fat class increases, the percentage of saleable meat decreases. The five conformation classes (EUROP) are based on the thickness of the flesh in relation to the skeleton, judged visually by blockiness and fullness of legs. The range is from E for extra to P for poor. The target area for most producers is 2–3L E, U or R.

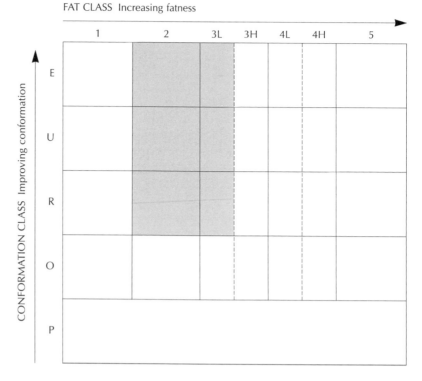

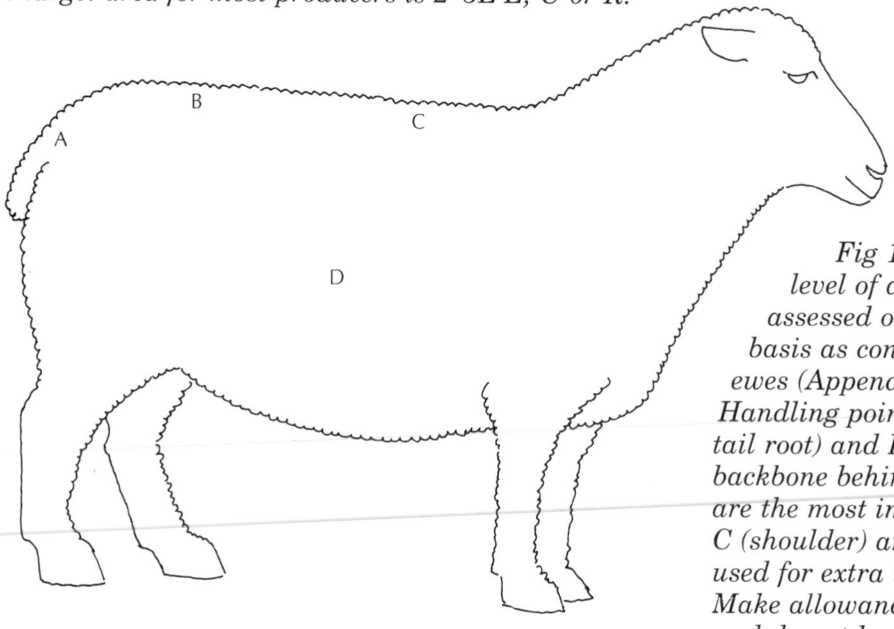

FAT CLASS Increasing fatness

CONFORMATION CLASS Improving conformation

	1	2	3L	3H	4L	4H	5
E							
U							
R							
O							
P							

Fig 106 The fat level of a lamb is assessed on the same basis as condition scoring ewes (Appendix II). Handling points A (dock or tail root) and B (the backbone behind the last rib) are the most important, with C (shoulder) and D (ribs) used for extra information. Make allowances for the wool and do not bruise by pressing too hard.

*Fig 107
Handling a lamb
to assess its
condition for
slaughter. Do not
press too hard
because the skin
may bruise.*

Check its accuracy by weighing an object of known weight such as a bag of feed and make sure that it is sited so as to work freely.

Set the weigher in a race so that lambs will be keen to escape through it. Unless there is system for separating the ewes and lambs, the ewes will have to go through the weigher as well or be physically held back.

At weighing, each lamb can be marked with a small spot either on different parts of the back or in different colours to denote its weight. This gives a visual indication in the field of how close they are to selling and allows lighter ones to be ignored in alternate weeks to speed up the weighing. Record all weights regularly and work out the growth rate of individuals and the average for the flock for the records and to estimate when lambs will be ready for market. Try to weigh at regular intervals and avoid wet weather when wet fleeces increase the weights.

Lambs which are ready to sell can be marked on the back of the neck and returned to the flock. Even if they have tags, a mark is quicker to see when they are finally mobbed and drafted. A mark on the top of the head is sometimes advocated but if they fight they can mark the head of another lamb and cause confusion.

When lambs are ready for market they should not be kept in the hope of getting better prices; they will get over-fat or lose condition, depending on the time of year.

BUSINESS

Any flock will involve a degree of book work, some of it required by law and some to monitor and improve the flock.

Records that must be kept by law in Europe include Movement Records and Medicine Records. The Movement Records

153

control sheep diseases in the national population by allowing the movements of all animals to be traced.

Animals being moved must be identified by a mark or tag and must be accompanied by a document to specify:

- Date of movement.
- Name of owner.
- Identification mark.
- Total number being moved.
- The holding from which they are being moved.
- The holding or premises to which they are being moved.

Medicine Records are intended to ensure that drugs, particularly antibiotics and those requiring withdrawal periods, are correctly used. Both record books must be kept up to date and both will be periodically checked.

Because of these controls on animals, flockowners must register their farm and their flock with local Ministry or Departments of Agriculture.

Records for Premium Schemes

Records may also have to be kept to verify claims under EU sheep premium payment schemes.

A number of schemes have been developed to support the sheep industry in Europe and operating currently is the Sheep Annual Premium Scheme (SAPS). A payment is made on each breeding ewe in the flock and the level of payment is linked to lamb prices in order to compensate producers for low lamb prices. The amount is calculated from the extent to which the average price of sheepmeat during the year falls below a European Union (EU) target price.

To claim the ewe premium producers must have quota – one quota per head of ewe. This quota may be allocated from a national reserve or bought or leased from another producer, either direct or through a broker. Quota was introduced to control the amount of subsidy being paid to the EC sheep industry and not to control production. This means that producers can still keep and sell sheep even if they have no quota; they will simply be ineligible for ewe premium payments. In the UK the scheme is administered by the government agricultural departments of MAFF (England), SOAFD (Scotland) and WOAD (Wales).

Under the SAPS a supplement is paid on ewes in Less Favoured Areas (LFA) – these are generally hill areas where farming is difficult. Because of the supplement, LFA quota and non-LFA (lowland) quota are 'ring fenced' so that quota cannot be bought and sold between the two areas.

Hill farmers in the UK also get financial help in the form of ewe headage payments under the Hill Livestock Compensatory Allowances (HLCA) and there are other systems of support in other EU countries.

Sheep farmers may also have access to various grant schemes to conserve the countryside, convert to organic production or run the flock at low stocking rates and enhance and conserve the landscape, wildlife and history. See Useful Addresses for some contacts.

Identification

Most record keeping is dependent on identifying individual sheep, so a reliable identification system is crucial. Ewes and lambs can be permanently marked either by a tag in the ear, a tattoo inside the ear, a number branded on the horn or an electronic microchip implanted under the skin. Plastic tags (Fig 108) are the easiest system for most adult flocks but do not always stay in place for the lifetime of the animal and are regarded as semi-permanent; permanent systems such as a tattoo are usually mandatory for pedigree breeders.

Demand for traceability in the meat industry may result in EC regulations making it compulsory for all sheep to carry a flock mark and an individual number so that they can be traced to their birth farm.

Fig 108 Plastic tags are the easiest way to identify and record individual sheep. Under EU traceability plans, types of tags and the information they carry may be subject to laws. Agricultural merchants sell a range of types. They must be applied hygienically, avoiding blood vessels. A small flock of docile sheep need not have large tags; but flighty ewes, which may need to be identified from a distance when a shepherd sees a problem, need big tags so that he knows which sheep to treat when he has them penned.

RECORDING

Records are essential for identifying any weakness in a flock or its management. Areas to record include:

- Physical performance (lambing percentage, wool weights, milk production).
- Financial performance (inputs, outputs, gross margins).
- Management records (tupping, housing, shearing dates and field uses).
- Sales of products (quantities, dates, prices and outlets).
- Individual ewes (breeding, health, production, wool weights).
- Health (flock and individual treatments).

Many of these can be combined and the information they contain will be tailored to the needs of the flock. Simple records are adequate but it is important to keep the information in the same format each year to make it easy to spot trends and make comparisons. They can also include target figures.

A record of flock lambing performance is a good example:

Ewes put to ram	25
Ewes died	1
Ewes culled	1
Ewes aborted	0
Ewes barren	1
Ewes lambed	22
Lambs born alive	44
Mean litter size	2
Lambs died	3
Number of lambs reared	41
Lambing percentage	164
Lambs retained for breeding	6
Lambs sold for slaughter	35
Average carcass weight	18.2kg

These figures indicate where improvements may be made such as increasing the number of ewes that lamb and the number of lambs that survive.

This example demonstrates that although the mean litter size was two, the flock did not achieve the magic 200 per cent lambing rate that many lowland flocks aspire to. The genuine lambing percentage (number of lambs reared per ewes put to the ram) was 164. The mean litter size or prolificacy (the number of lambs born per ewes lambing) indicates whether or not the ewes that have lambed are reaching their breed potential.

Financial Performance

The small flock is unlikely to make much profit unless it hits the big time in pedigree

Table 9 Example Gross Margin Calculation

Income (Per ewe put to the ram)

Sales of lamb/valuation	75.00
Wool	3.00
Ewe premium	10.00
Total returns	*88.00*

Variable costs

Flock replacements	10.00
(ewes and rams)	
Ewe concentrates	10.00
Lamb concentrates	5.00
Other purchased feed	3.00
Forage/grassland costs	8.00
Vet/medicines	5.00
Market/transport	2.00
Sundries	2.00
Total variable costs	*45.00*
Gross margin per ewe	43.00
Stocking rate (grass and	
forage) ewes per hectare	8
Gross margin per hectare	*344.00*

The figures used in this table are for illustration purposes only.

of one enterprise with that of another. Gross margin calculations could demonstrate whether or not it is better to sell the sheep and use the land for growing vegetables.

Accurate gross margins require an accurate record of income and costs.

Table 9 is an example of a gross margin calculation. The figures are for illustration only.

The figures highlight some of the areas where improvements may be made, but improvements in one area may be countered by added costs:

- Rearing more lambs. (Improved management or a more prolific breed. But high lambing percentages can be counter-productive by producing small lambs with high mortality rates and the expense of rearing surplus lambs.)
- Higher stocking rate. (As this increases, individual performance may fall as sheep compete for food. There is a compromise between performance per hectare and performance per animal.)
- Higher lamb return per ewe. (Better breeding, better marketing.)
- Lower feed costs. (Better use of grassland.)

Management Records

A record of the key dates in the flock calendar also form the basis of a blueprint for planning the year ahead.

- Flushing
- Tupping
- Start and finish of supplementary feeding
- Housing
- Lambing
- Routine veterinary treatments
  (Worming, vaccination, foot care)
- Weaning
- Shearing
- Dates and results of condition scoring
- Stock sale dates
- Wool sale dates

breed prices or adds value to its products. But most small flocks aspire to being large flocks, and a record of financial performances highlights areas for improvement.

Gross margins are a traditional way of expressing the financial performance of an enterprise. The gross margin figure is simply the output of a flock less the variable costs. This is not the profit because it does not include fixed costs such as labour, rent, insurance and depreciation.

The value of gross margins is to allow a comparison of financial performance from year to year and to compare the performance

Sales of Products

A record of slaughter lamb sales will monitor and analyse the marketing side of the project and can be compared with target figures.

		Target
Draft number	3rd	
Date drafted	25 March	
Date sold	2 April	19 March
Where sold	Ashton Meat	
Number sold	10	15
Average carcass wt	17.7kg	18kg
Average price/kg dw	254p	270p
Average price per head	£45	£48
Average classification figures	3LR	3E/3U
Comments	Very mixed weights One fat class 5	

Individual Ewe Records

Records for individual breeding animals are important in commercial as well as pedigree flocks. They monitor performance to identify the best and the worst ewes and include health and lambing details. Such records can identify families that may have an hereditary health problem. A card file is adequate and is an easy reference.

Information to be recorded includes:

Identification of ewe	Red Tag 11	
Sire and dam	Suffolk Ben × Masham 108	
Date of birth	February 2 1996	
Litter size	3	
Reared as	Twin	
Birth weight	4kg	
Weaning weight	32kg	
Fleece weight	Year 1	3kg
	Year 2	3.5kg

Veterinary record

Date	Aug 1996	Aug 1997
Feet	Good	Good
Udder	Small teats	Good
Teeth	Temp.	2
Cond. score	3.5	3.0

Veterinary comments
12 June 1998   Treatment for orf

Lambing record

Date	16/2	
Service	1st	
CS	2.5	
Litter	2	
Reared	2	
Sexes	E	R
Sire	Suff	Suff
Assistance	None	Breech
Birth Weight	3.5kg	4.1kg
Wean Weight	33kg	36kg
Sales	Retained	24/5 18kg
Notes	Nervous. Slow to milk	

Disposal

Date	5/9/98
Reason	Surplus
Method	Private sale to local flock

Flock recording schemes are run by Signet and independent consultants. They monitor and advise on breeding, production and financial aspects of a flock. Producers who sell breeding stock can have independently-monitored performance figures to show to prospective purchasers.

For computer buffs there are flock recording programmes available and an advance in electronic tagging means that a microchip implant in the sheep's ear can store information about the individual that can be logged on a hand-held reader for downloading to a computer.

12 The Shepherd's Year

Sheep production needs a routine. Small flocks inevitably have part-time shepherding and are fitted in with other enterprises or work, so forward planning is essential to:

- Avoid forgetting.
- Be timely.
- Avoid clashing with other activities.
- Keep production on schedule to meet the markets.

The easiest approach is to draw up a blueprint for routine tasks. The following calendar is basic and flock owners should draw up their own, adding extra reminders as appropriate.

Among these should be grassland management dates such as:

- Topdressing.
- Closing dates for hay or silage fields.
- Cultivation and reseeding.
- Sowing and harvesting root or forage crops.
- Soil analyses.

Especially important are dates for ordering and buying goods such as lambing equipment, raddle markers, fertilizers, grass seeds, fencing posts, vaccines, wormers, and supplementary feeds, and so on. Also, when to buy replacement stock such as rams and when to fill in and return grant and premium application forms.

Month 1
Preparation for mating
Check teeth, feet and udders.

In Europe this calendar is adaptable to various lambing systems as shown in the table.		
System	**Lambing**	**Month 1 (Start of cycle)**
Autumn lambing	October	April
Early lambing	January	July
Spring lambing	March	September
Late lambing	May	November

Clostridial and pasteurella booster vaccination.
Worm.
Crutch ewes.
Separate into tupping groups.
Flush on improved grass/supplementary feed.
Have marking crayons and raddle or paste ready.

Where appropriate
Vaccinate against abortion.
Dip.
Footrot vaccine.
Start a dog worming programme for the year.
Synchronize by teaser or sponges.

Month 2
Tupping
Put in rams with raddle/paste.
Maintain level of nutrition of ewes.
Hand feed rams to maintain condition if working hard.

Change raddle/paste colour every ten to seventeen days.
Do not disturb the flock.
Note dates as ewes are marked.
Watch that rams working and not lame.
Beware of high number of returns – suggesting infertile ram.

If appropriate
Move rams around tupping groups.
Check harness in place and not chafing.

Month 3
Implantation
Maintain level of nutrition. Do not overfeed.
Keep eye on feet and general condition.
Do not disturb.
Check harness in place and not chafing.

Month 4
Early pregnancy
Remove rams (after five to six weeks).
Maintain nutrition but do not overfeed.
Check condition scores and adjust feed.

If appropriate
Prepare housing.
Vaccinate against footrot.

Month 5
Mid-pregnancy
Pregnancy scan ewes (twelve weeks after rams go in for mating).
Repair fence/hedges.
Introduce supplementary feed.
Condition score and adjust feed of individuals if necessary.

If appropriate
House.
Shear or crutch.

Month 6
Late pregnancy
Watch for metabolic problems and abortion.
Watch for footrot and lice in housed ewes.
Prepare for lambing.
Increase supplementary feed.

Prepare grassland or forage area for lambed flock.
Clostridial booster (vaccinate first and later lambers separately if prolonged lambing to give better protection to lambs).

If appropriate
House.
Worm at housing or wait until post-lambing.

Month 7
Lambing
Prepare for turnout into sheltered fields.
Improve nutrition of lactating ewes especially if on short pasture.
Consider adding magnesium to rations if hypomagnesaemia is a problem.
Watch for mastitis, hypothermia, starvation.
Beware predators.

If appropriate
Creep feed lambs.
Worm ewes at turnout.
Check feet.

Month 8
Rearing
Clean out and disinfect lambing shed/area.
Watch for nematodirus in lambs.
Start regular drenching if not on clean grazing.
Watch for coccidiosis.
Beware cobalt and other trace element deficiencies.

If appropriate
Creep feed lambs.
Continue to supplement ewes.

Month 9
Rearing
Continue regular worm drenching if necessary.
Watch for fly strike and headfly in ewes and lambs.
Start weighing lambs to monitor growth.
Beware early born singles that are ready for slaughter.

Month 10
Rearing and selling
Weigh lambs regularly and handle for finish.
Identify ewe lamb replacements.
Protect against fly strike.

If appropriate
Plan shearing.
Plan dipping.
Continue worming.

Month 11
Weaning and selling
Have clean grazing area for lambs. Lambs on best grazing.
Ewes on poor grazing.
Watch for mastitis. Consider dry ewe therapy.
Start clostridial vaccinations for ewe lambs and for finishing lambs if likely to be kept beyond four months.
Draw lambs weekly.
Identify and sell or fatten cull ewes.

Make sure good grass/forage will be available for flushing.

If appropriate
Have a blitz against footrot.
Worm lambs regularly but not if close to sale.

Month 12
Dry
Condition score ewes and adjust feed appropriately.
Put ewe lamb replacements in their own flock.
Give ewe lambs permanent identification.
Keep ewe lamb replacements growing steadily.
Prepare rams for tupping.

If appropriate
Shear and dip. There should be at least three weeks' wool growth at dipping.
Buy in ewe or ram replacements. Worm, vaccinate and isolate for a month.

Appendix I
Anatomy of the Sheep

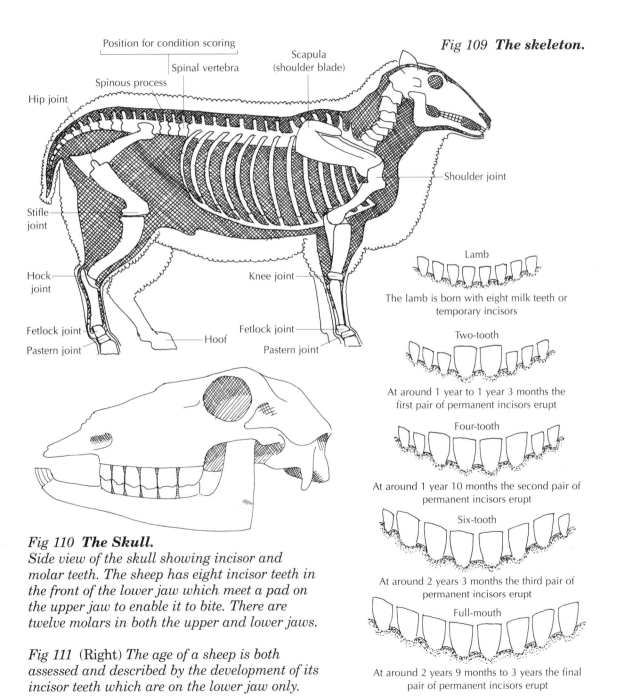

Position for condition scoring

Spinal vertebra

Scapula (shoulder blade)

Fig 109 **The skeleton.**

Spinous process

Hip joint

Shoulder joint

Stifle joint

Hock joint

Knee joint

Fetlock joint

Hoof

Pastern joint

Fetlock joint

Pastern joint

Lamb

The lamb is born with eight milk teeth or temporary incisors

Two-tooth

At around 1 year to 1 year 3 months the first pair of permanent incisors erupt

Four-tooth

At around 1 year 10 months the second pair of permanent incisors erupt

Six-tooth

At around 2 years 3 months the third pair of permanent incisors erupt

Full-mouth

At around 2 years 9 months to 3 years the final pair of permanent incisors erupt

Fig 110 **The Skull.**
Side view of the skull showing incisor and molar teeth. The sheep has eight incisor teeth in the front of the lower jaw which meet a pad on the upper jaw to enable it to bite. There are twelve molars in both the upper and lower jaws.

Fig 111 (Right) *The age of a sheep is both assessed and described by the development of its incisor teeth which are on the lower jaw only.*

161

Fig 112 **The external points.**

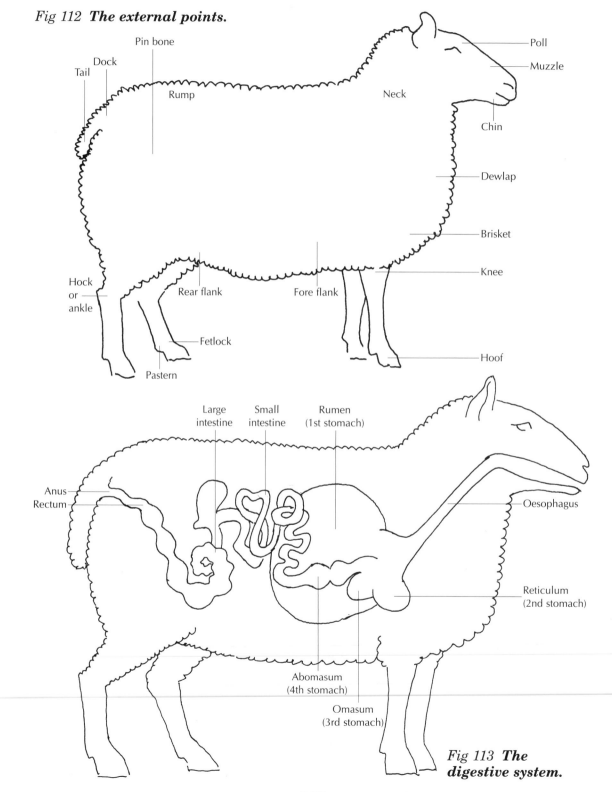

Fig 113 **The digestive system.**

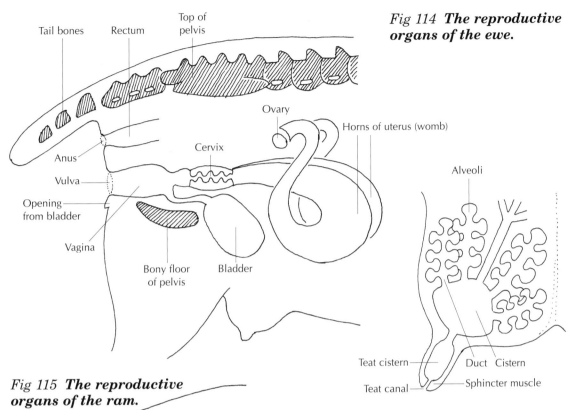

Fig 114 *The reproductive organs of the ewe.*

Tail bones
Rectum
Top of pelvis
Anus
Vulva
Opening from bladder
Vagina
Bony floor of pelvis
Cervix
Bladder
Ovary
Horns of uterus (womb)

Alveoli
Teat cistern
Teat canal
Duct Cistern
Sphincter muscle

Fig 115 *The reproductive organs of the ram.*

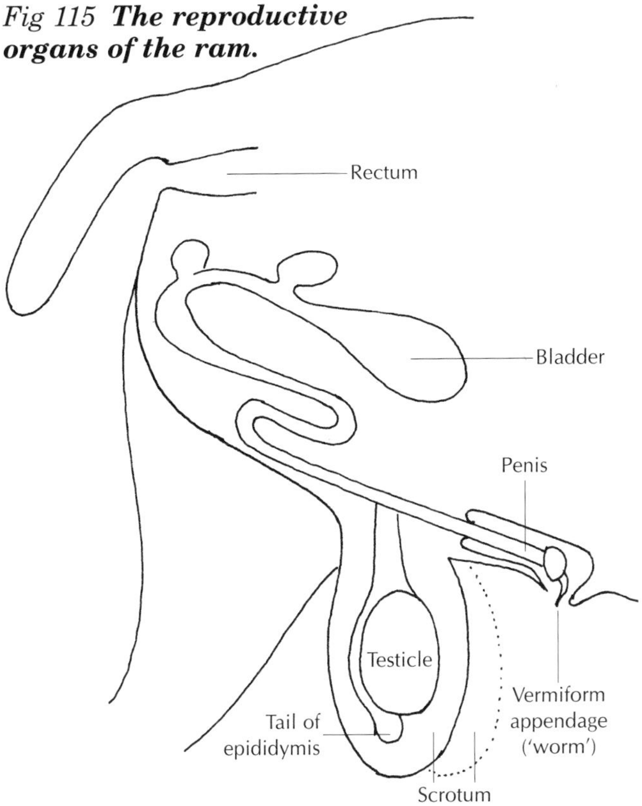

Rectum
Bladder
Penis
Testicle
Tail of epididymis
Vermiform appendage ('worm')
Scrotum

Fig 116 *The udder. The ewe's udder is in two separate halves, each with a teat. A ewe can lose the use of one half (usually called a 'quarter') from mastitis or injury but continue to milk in the other half. Cells in the alveoli produce milk which goes via the ducts to the cistern, which acts as a reservoir. Milk production is continuous and the rate is controlled by pressure, so that when a ewe is weaned and the pressure of the milk builds up in the udder the rate of production slows down and stops. The sphincter muscle closes the teat against leaking and invasion by bacteria. After weaning, a plug forms in the teat canal to prevent the entry of dirt and disease, so it is often better not to strip a ewe (milk it by hand) after weaning even if her udder looks uncomfortable.*

Appendix II

Condition Scoring (and Lamb Classification)

Condition scoring measures the fatness of an animal and describes its condition. It is one of the shepherd's best management tools.

The system was devised in Australia and is used to monitor the condition of ewes, especially in relation to feeding. It is important to practise condition scoring regularly to get the feel for it; even if it is not very accurate it will give an indication of any changes in the condition of a sheep. With practice the condition can be assessed in half scores. The system is less effective with some rare and primitive sheep breeds which do not lay down external fat.

The score is assessed by finger pressure in the lumbar region (the backbone immediately behind the last rib – Appendix I, Fig 109) where condition is laid down first and lost first. Assess in order:

A. Prominence (sharpness or roundness) of the spinous processes.
B. Prominence of the transverse processes.

C. Amount of muscle and fat tissue under the transverse processes by the ease with which the fingers pass under the bones.
D. Fullness of the eye muscle and its fat cover between the spinous and transverse processes.

Assessing lambs for slaughter
The system is also the basis for selecting lambs for slaughter and describing the finish of carcasses (classification).

Fat class 1: Fat cover very thin and bones easy to feel
Fat class 2: Fat cover thin and bones easy to feel with light pressure
Fat class 3: Individual bones felt with light pressure
Fat class 4: Fat cover quite thick and individual bones felt only with pressure.
Fat class 5: Fat cover thick and bones cannot be felt.

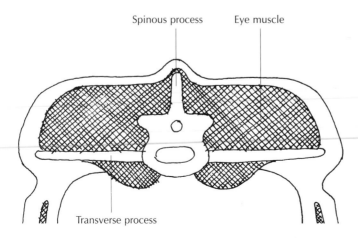

Spinous process   Eye muscle

Transverse process

*Fig 117 **Score 1: very lean.** Spinous and transverse processes are sharp and fingers pass easily under the ends. The eye muscle is shallow with no fat cover. Looks thin.*

Fig 118 **Score 2: lean.**
The spinous processes are prominent but smooth and feel like corrugations. The transverse processes are smooth and rounded. Fingers pass under the ends with a little pressure. Eye muscle depth is moderate but there is no fat.

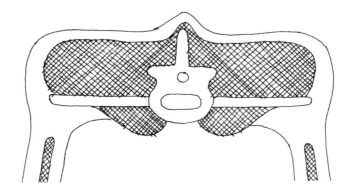

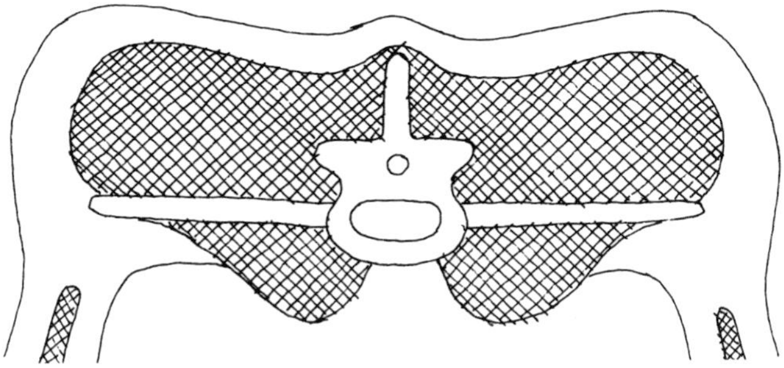

Fig 119 **Score 3: good condition.**
The spinous processes have small elevations that are smooth and rounded and individual bones can be felt with pressure. The transverse processes are well covered and pressure is needed to feel the ends. Eye muscle is full with some fat cover.

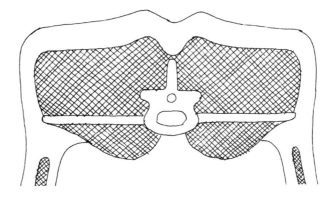

Fig 120 **Score 4: fat.**
It takes pressure to feel the spinous processes and the ends of the transverse processes cannot be felt. The eye muscle is full and has a thick fat cover.

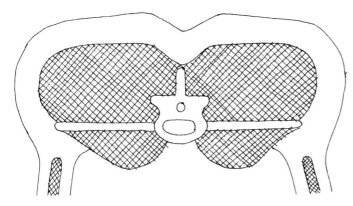

Fig 121 **Score 5: very fat.**
Neither the spinous processes nor the transverse processes can be felt. The eye muscle is full with thick fat cover and there may be thick fat cover on the rump and tail. The animal will look fat.

Glossary

Abortion – premature birth of a lamb that cannot survive outside the ewe.

Accredited flocks – ones which have been officially cleared of specific diseases.

Acidosis – acidity in the rumen usually caused by overeating cereals.

Acre – 0.4 hectare.

Aftermath – grass regrowth after cutting for hay or silage.

Anaerobic – without oxygen.

Anthelmintic – the drug used to kill worms/internal parasites/helminths.

Antibodies – proteins produced in the blood to fight disease.

Artificial insemination – semen is collected from a ram and put into the reproductive tract of a ewe.

Barrener – ewe not in lamb.

Blind quarter – one half of the udder that is not functioning.

Break – a weakness in the fleece usually caused by poor nutrition or stress.

Break crop – a crop grown for a season to interrupt grass or another crop to break a cycle of weeds and disease.

Breed – a population of sheep that have obviously different characteristics from other sheep and that are genetically determined.

Broadcast – sowing seeds from above the surface as from a fertilizer spreader rather than drilling into the ground.

Broken mouth – an adult sheep that has lost some teeth.

Browse – eat the shoots and leaves of trees and shrubs.

Cade lamb – a hand-reared lamb.

Carrying capacity – the number of sheep that can live off a given area of land.

Cast ewe – Cull ewe. Also ewe stuck on its back.

Clean grazing – grazing area free from internal parasites.

Cleats – two halves of the hoof.

Colostrum – first milk produced after lambing.

Compensatory growth – after a period of food shortage when growth is faster than would be expected from the diet.

Compounds – concentrates processed into pellet form by compounders.

Concentrates – High-energy, high-protein feed to supplement forage.

Condition – level of fatness.

Conformation – shape and muscling of the body.

Coronet – top of the hoof.

Cotted – fleece that is matted or felted.

Couples – ewes and lambs.

Creep feed – compound feed given to lambs before weaning.

Creep – special feeding or grazing area for lambs, inaccessible to the ewes.

Crimp – waviness in wool.

Crossbred – sheep with parents of two different breeds.

Crossing sire – ram used to sire breeding ewes.

Crutching – shearing wool from around the tail area of a sheep to keep them clean, and to aid mating and lambing.

Cuckoo lamb – born in early summer. Usually a late lamb.

Cud – food regurgitated for chewing.

Cull – a sheep removed from a flock because it is no longer wanted for breeding.

Dagging – trimming faeces from the wool.

Dags – dung attached to the fleece, usually around the tail.

Deadweight selling – lambs sold direct to abattoirs; the producer is paid for the carcass. Also called selling 'on the hook'.

Docking – removing the tail.

Dosing – giving a dose of medicine orally. Also called drenching.

Double – a twin.

Draft ewe – breeding ewe drafted from a hill flock for breeding on a lowland farm.

Drafting – separating sheep from the flock. Such as drafting lambs from their mothers at weaning or drafting sheep for sale.

Drawing – selecting lambs for slaughter. Also pulling a lamb from a ewe at birth.

Drenching – giving a dose of medicine orally. Also called dosing.

Dry ewes – ewes which are not lactating.

Drying-off – ewes drying up their milk at the end of lactation.

Dry matter – the component of feed left when all moisture is removed.

Dry sheep – sheep kept without breeding from them, usually for wool production.

Dystokia – difficult birth.

Electrolyte – a liquid that corrects dehydration in the body.

Entire – a male sheep that is not castrated.

Eruption – adult teeth coming through the gum.

Ewe – adult female sheep.

Fat lamb – Once used to describe a slaughter lamb but now politically incorrect because fat is not wanted. Now called prime lamb.

Finish – The level of fat on a live sheep or carcass.

Finishing – getting a lamb to the correct weight and fat level suitable for selling for slaughter. The old term 'fattening' is politically incorrect.

Flushing – Improving nutrition prior to mating.

Foetus – a lamb before it is born.

FOM – a lamb sold finished off its mother before weaning.

Forage – any plant material (except that used in concentrates) used as a food for sheep and other herbivores. Fibrous crops for ruminants.

Foster lamb – one reared by a ewe that is not its natural mother.

Gestation period – duration or length of pregnancy from conception to birth.

Gimmer – a young female sheep.

Grading – judging the fat and conformation of a slaughter lamb.

Grassland – a plant community dominated by grasses and clovers.

Graze – to eat growing vegetation close to the ground.

Half-bred – traditionally a sheep sired by the Border Leicester.

Hay – grass preserved by drying for feeding in the winter.

Heading date – when the ear or head of a grass plant emerges. For a sward, the heading date is when 50 per cent of the ears have emerged.

Hectare – 2.5 acres.

Helminth – parasitic worm.

Hermaphrodite – animal with the characteristics of both sexes.

Hogget – heavy lamb sold at 8–12 months of age and after 1 January (UK). Also, between six months old and being shorn for the first time.

Hybrid vigour (heterosis) – improved performance of a crossbred over the performance of its parents.

Hyperthermia – higher than normal body temperature.

Hypothermia – lower than normal body temperature.

Inbreeding – the breeding of closely related animals.

In-bye – the best grazing on a hill farm, usually at lower altitudes.

Intramuscular – in the muscles.

Judas sheep – a tame sheep which follows the shepherd and leads the flock.

Kemp – Coarse hair in a fleece.

Killing out percentage – percentage of the live lamb that ends up as a carcass.

Lactation – period when milk is produced and secreted from the udder.

Lambing percentage – number of lambs reared per one hundred ewes put to rams.

Ley – land temporarily under grass.

Liveweight selling – stock are sold through livestock markets or similar and are paid for on their live weight. Also known as 'on the hoof'.

Maintenance feed – a level of feeding to maintain body weight and health but not production.

Meconium – the first faeces passed by a lamb.

Metabolizable energy – the energy in feed that is used by the sheep.

Mismothering – lambs attaching themselves to the wrong ewe.

Mob – to get the sheep to flock together in a tight group.

Moiety – straw or other vegetable contamination found in fleeces.

Mothering-on – bonding a ewe and lamb.

Mule – traditionally a sheep sired by the Bluefaced Leicester.

Notch – an identification mark cut in the edge of the ear.

Nursing – a ewe which is suckling a lamb.

Oestrus – period when ewes are ready to mate.

Pathogen – an agent causing disease.

Periparturient rise – when internal parasites in ewes shed excessive numbers of eggs at around lambing time. Also known as the 'spring rise'.

Permanent pasture – long-term grassland that is rarely ploughed.

pH – a measure of the acidity of soil.

Pheromone – chemical substance that produces an odour to attract a mate.

Poaching – damage to soil and crops by animals' feet especially in wet weather.

Polled – hornless.

Pour-on – a veterinary treatment that is applied by pouring or spraying along the back of a sheep.

Prime lamb – first-class slaughter lamb. Used in preference to fat lamb.

Progeny – immediate descendants of a sheep.

Prolific – having large litters of lambs. A ewe who averages litters of three lambs or more would be described as prolific.

Quarter – one half of a ewe's udder.

Race – a narrow pen that sheep can walk through in single file for treatment.

Ram – adult male sheep.

Replacement – a sheep bought or bred to replace another culled from the flock.

Roughage – fibrous food (grass, hay, silage, straw) necessary for the function of the rumen.

Rumen – largest of the ruminant's four stomachs.

Ruminant – an animal that chews the cud.

Scanning – ultra sound scanning to identify the number of lambs in a pregnant ewe. Will also measure the depth of fat and muscle.

Scouring – having diarrhoea.

Selection – choosing sheep as parents of the next generation.

Shearling – young sheep between its first and second shearing.

Silage – grass preserved by 'pickling' for feeding in the winter.

Spring rise – see periparturient rise.

Staple – 'bundles' of wool fibres in a fleece.

Still-born – A lamb born dead and nothing can be done to save it.

Stocking rate – number of adult sheep that a hectare of land can support for a year.

Store lambs – unfinished lambs bought to finish later in the year. Those unfinished by 1 October (UK).

Strike – when affected by blowfly maggots. Fly strike.

Strip grazing – fencing pasture or forage crop temporarily so that animals graze a strip at a time.

Stripping – milking a ewe by hand.

Strong wool – coarse wool.

Subcutaneous – under the skin.

Supplement – food and minerals given in addition to the main diet.

Supplementary feed – any feed given to supplement the basal diet.

Sward – an area of grassland with a short, continuous cover of foliage.

Symptom – a change seen as a result of disease.

Synchronized – managed so that the flock will mate (and lamb) over a short period.

Tag – a form of identification put in the ear.

Terminal sire – ram which sires lambs for slaughter.

Tillering – vegetative reproduction of grasses when they send out new side shoots from their base.

Topdressing – spreading fertilizer.

Topping – cutting off the seedheads and weeds in grassland.

Topping-up – giving extra milk from a bottle to a sucking lamb.

Toxic – poisonous.

Tup – a ram.

Tupping – mating.

Turnout – when ewes are turned out to grass after winter housing.

Weaning – when a lamb no longer needs milk and is removed from its mother.

Wether – castrated ram lamb.

Windchill – loss of body heat caused by wind and rain.

Winter kill – death of plants in the winter.

Yeld or eild – barren ewe.

Yolk or suint – yellow substance secreted by the skin into the fleece.

Zero grazing – cutting fresh grass and taking it to the animals.

Zoonoses – disease that is transmissible between sheep and man.

Further Reading

Black's Veterinary Dictionary,
Ed. Geoffrey West.
A & C Black.
A comprehensive A–Z of every animal ailment and veterinary jargon.

British Sheep,
National Sheep Association.
All the breeds of sheep plus facts about the industry.

Feeding the Ewe,
Meat and Livestock Commission.
The authority on the nutritional needs of sheep.

Handbook for the Sheep Clinician,
M Clarkson.
Liverpool University Press.
A practical approach to sheep care from the veterinary surgeon's view.

Improved Grassland Management,
John Frame.
Farming Press Books.
All about grassland. Very informative and very readable.

Modern Shepherd,
Brown and Meadowcroft.
Farming Press Books.
Aimed at the big flock professional, but all the facts are there.

Planned Sheep Production,
David Croston and Geoff Pollott.
Blackwell Scientific Publications.
An overview of the sheep industry. Very factual.

Practical Lambing,
Eales and Small.
Longman Scientific and Technical.
Buy this before you buy the sheep. Should be compulsory reading for all flockowners.

Practical Sheep Dairying,
Olivia Mills.
Thorsons Publishing Group.
The authority on sheep dairying.

Sheep Ailments,
Eddie Straiton.
Farming Press Books.
A classic. The pictures say it all.

The Showman Shepherd,
David Turner.
Farming Press Books
Practical and friendly advice for the aspiring showman.

The Veterinary Book for Sheep Farmers,
David Henderson.
Farming Press Books.
Incredibly comprehensive. Relates the veterinary side of sheep keeping to management and explains everything.

Useful Addresses

ADAS Consultancy Ltd
ADAS Headquarters, Oxford Spires
Business Park, Kidlington, Oxon OX5 1NZ.
Consultants and advisors on all aspects of
sheep farming.

Agricultural Training Board
Stoneleigh Park Pavilion, NAC,
Kenilworth, Warks CV8 2VG.
Agricultural training schemes and advice.

**British Association of
Sheep Contractors**
Cherry Croft Farm Cottage, Compton,
Winchester, Hants.
Represents sheep contractors.

British Grassland Society
No 1 Earley Gate, University of Reading,
Reading RG6 6AT UK.
Involved in research and practical
grassland production. Local branches.

British Organic Farmers
86 Colston St., Bristol BS1 5BB.

British Sheep Dairying Association
Wield Wood, Alresford, Hampshire SO24
9RU.
Information on milking sheep.

British Wool Marketing Board
Oak Mills, Station Road, Clayton,
Bradford, West Yorkshire, BD14 6JD.
Markets wool on behalf of producers.
Shearing courses, advice on wool
production and handling, fleece
assessments.

**Department of Agriculture for
Northern Ireland** (DANI)
Dundonald House, Upper Newtownards
Road, Belfast BT4 3SB.
Advice for sheep farmers.

Environment Agency
Rio House, Waterside Drive,
Aztec West, Almondsbury, Bristol
BS12 4UD (Local offices in phone book).
Advice on waste disposal such as sheep
dip and carcasses and regulations con-
cerning the use of streams and rivers.

**Farming and Wildlife
Advisory Group** (FWAG)
National Agricultural Centre, Stoneleigh,
Kenilworth, Warwickshire CV8 2RX.
Advice on shelter belts and other practical
conservation issues.

Farmers' Union of Wales
Llys Amaeth, Queen's Square,
Aberystwyth, Dyfed SY23 2AE.
Represents interests of Welsh farmers.

Health and Safety Executive
Broad Lane, Sheffield, S3 7HQ
(Local offices in phone book).
Advice on health and safety in agriculture.

Humane Slaughter Association (HSA)
The Old School House, Brewhouse Hill,
Wheathampstead, Hertfordshire AL4 8AN.
Advice on the handling of slaughter
animals and casualty slaughter.

International Sheepdog Society
Chesham House, 47 Bromham Road,
Bedford MK40.

Meat and Livestock Commission
PO Box 44, Winterhill House, Snowdon
Drive, Milton Keynes, MK6 1AX.
Funds research into sheep production
and marketing. Administers Signet.

**Ministry of Agriculture, Fisheries
and Food** (MAFF)
3 Whitehall Place, London SW1A 2HH.
See phone book for Regional Service
Centres that are contact points for
Ministry schemes.

The Moredun Foundation
Pentlands Science Park, Bush Loan,
Penicuik, Midlothian, Scotland.
Research in sheep health. Findings
available to members.

National Farmers' Union
164 Shaftesbury Avenue,
London WC2H 8HL.
Represents the interests of English
farmers.

National Farmers' Union of Scotland
Rural Centre, West Mains, Ingliston,
Newbridge, Midlothian EH28 8LT.
Represents the interests of Scottish
farmers.

National Sheep Association
The Sheep Centre, Malvern,
Worcs WR13 6PH.
Represents the interests of sheep farmers.
Information to members through the
Sheep Farmer magazine. Source of
information on breed societies, legislation,
and so on.

Rare Breeds Survival Trust
National Agricultural Centre,
Kenilworth, Warks CV8 2LG.
Represents the interests of rare breeds.
Information to members through *The Ark*
magazine.

**Royal Society for the Prevention
of Cruelty to Animals** (RSPCA)
The Causeway, Horsham, West Sussex
RH12 1HG. (Local offices in phone book).
Advice on sheep welfare and the law.

SAC (The Scottish Agricultural College)
SAC Central Office, West Mains Road,
Edinburgh EH9 3JG.
Research and advisory service. Runs
national sheep health schemes.

**The Scottish Office Agriculture and
Fisheries Department** (SOAFD)
Pentland House, 47 Robbs Loan,
Edinburgh EH14 1TW.
Contact for Ministry schemes.

Signet
PO Box 603, Winterhill House, Snowdon
Drive, Milton Keynes MK6 1AX.
Flock improvement schemes, advice
on sheep production. Animal breeding
division of the Meat and Livestock
Commission.

Small Acres Advisory Service
Holly Lodge, Spencers Lane, Berkswell,
Coventry CV7 7BZ.
Independent specialist advice on small
flock management.

Teagasc
19 Sandymount Ave, Dublin 4, Ireland.
Practical advice on sheep management.

Ulster Farmers Union
475 Antrim Road, Belfast BT15 3DA.
Represents interests of farmers in Northern
Ireland.

Welsh Office Agricultural Department (WOAD)
Cathays Park, Cardiff CF1 3NQ.
Contact point for government schemes.

General Sources of Information
Commercial firms such as feed compounders, fertilizer manufacturers, seed merchants and veterinary and pharmaceutical companies; veterinary surgeons; agricultural merchants; land agents; market auctioneers; agricultural colleges.

Index ————————————————————